MODERN FLEXIBLE MULTI-BODY DYNAMICS MODELING METHODOLOGY FOR FLAPPING WING VEHICLES

MODERN FLEXIBLE MULTI-BODY DYNAMICS MODELING METHODOLOGY FOR FLAPPING WING VEHICLES

CORNELIA ALTENBUCHNER
NASA Jet Propulsion Laboratory, Pasadena, CA, USA
JAMES E. HUBBARD JR.
University of Maryland, Way Hampton, VA, USA

ELSEVIER

ACADEMIC PRESS
An imprint of Elsevier

Academic Press is an imprint of Elsevier
125 London Wall, London EC2Y 5AS, United Kingdom
525 B Street, Suite 1800, San Diego, CA 92101-4495, United States
50 Hampshire Street, 5th Floor, Cambridge, MA 02139, United States
The Boulevard, Langford Lane, Kidlington, Oxford OX5 1GB, United Kingdom

Library of Congress Cataloging-in-Publication Data
A catalog record for this book is available from the Library of Congress

British Library Cataloguing-in-Publication Data
A catalogue record for this book is available from the British Library

ISBN: 978-0-12-814136-6

For information on all Academic Press publications visit our website at
https://www.elsevier.com/books-and-journals

Publisher: Matthew Deans
Acquisition Editor: Carrie Bolger
Editorial Project Manager: Carrie Bolger
Production Project Manager: Paul Prasad Chandramohan
Designer: Victoria Pearson

Typeset by TNQ Books and Journals

The authors would like to dedicate this work to their beloved families and the beauty, elegance, and grace of nature.

CONTENTS

Preface *ix*
List of Figures *xi*
List of Tables *xvii*
List of Nomenclature *xix*
Acknowledgments *xxv*
Summary *xxvii*

1. Bioinspired Flight Robotics Systems **1**
1.1 Introduction of This Body of Work 1
1.2 The Background of Flapping Wing Flight Technology 1
1.3 A Model of an Ornithopter for Performance Optimization 4
1.4 Historical Considerations for Bioinspired Flapping Wings Avian Flight and Robotics 14
1.5 Objectives in the Development of Flexible Multi-Body Dynamics the Modeling Methodology Described in This Body of Work 18
References 19

2. Flexible Multi-Body Dynamics Modeling Methodology's for Flapping Wing Vehicles **23**
2.1 Classic Modeling Methodology's 23
2.2 Modern Modeling Methodology 44
References 47

3. Bioinspired Flapping Wing Test Platform Used to Implement Modern Modeling Methodology **51**
3.1 Details of the Test Platform 51
3.2 Experimental Data Sets of Bioinspired Flaping Wing Robotic System for Model Verification 57
References 72

4. Flexible Multi-Body Dynamics Modeling Methodology Implementation Avian Scale Flapping Wing Flyer **73**
4.1 Linear Elastic Multi-Body Systems 74
4.2 The Five-Body Multi-Body Dynamics Model 79
4.3 Relevant Coordinate Systems 80
4.4 An Underlying Articulated Rigid-Body Model 82

4.5 Lagrange Formulation of Equations of Motion 86
4.6 Formulation of Five-Body Flexible Multi-Body Dynamics Model 91
4.7 Structural Dynamics Model of the Wings 103
References 106

5. Aerodynamics Modeling for Flexible Multi-Body Dynamics Modeling Methodology Implementation Avian Scale Flapping Wing Flyer 109

5.1 The Aerodynamic Model Versions Formulated 109
5.2 Aero-Model A 112
5.3 Aero-Model B/C 113
5.4 The Aerodynamic Model Implementation 117
5.5 The Global Resulting Forces and Moments 124
References 127

6. Results of the Modeling Methodology Implementation and Flight Simulation 129

6.1 Modeling Assumptions Verification and Wing Flexibility 129
6.2 Model Results 143
6.3 Constraint Model Verification 144
6.4 Unconstraint Model Verification 147
References 153

7. Concluding Remarks About Modern Modeling Methodology Implementation and Flight Physics of Avian Scale Flight Robotics Systems 155

7.1 Summary of Modern Modeling Methodology Development and Implementation 155
7.2 Scope and Contributions Resulting From Modern Modeling Methodology Implementation Described in This Book 156
7.3 Summary of Novel Contributions Resulting From Modeling Methodology 159
7.4 Summary of Conclusions About the Modern Modeling Methodology 160
7.5 Summary of Recommendations for Modeling of Avian Scale Flapping Wing Flyers 161

Index *163*

PREFACE

During the past decade the Nation has seen a surge of interest in small unmanned vehicles. There has been a particular focus on bio-inspired flight as a design paradigm for these platforms. The application of biomimetic principles to the development of small unmanned vehicles spans the scale of flight from insects to avian flyers. Biomimetics is defined as the imitation of the models, systems, and elements of nature for the purpose of solving complex problems. Researchers, engineers, hobbyist, and aircraft designers all over the world now use nature to inspire and inform the design of modern unmanned platforms. There is no doubt that animal flight has played such a role for centuries, but recent times have seen an increased interest.

Avian-inspired birdlike robotic designs have given rise to modern ornithopters whose aerodynamic and kinematic properties are marginally predictable and can produce nonlinear behaviors, which are time varying and involve multiple dynamic scales. The goal of the work presented herein is to develop canonical modeling techniques that can drive the engineering design of avian scale flapping wing ornithopters.

Our approach to the modeling of such a complex vehicle is an energy-based Lagrangian approach that includes the flexible multi-body dynamics characteristic of this class of platforms. We are particularly interested in developing a model form and function suitable for feedback control. This will allow the work here to form the basis of algorithms, which can be implemented for improved agility, stability, lift, and thrust.

In developing the approach presented here the authors worked closely with NASA engineers, drone pilots, academic scholars, and designers to understand the subtleties of low Reynolds number flight and its associated dynamics and aerodynamics. We are particularly grateful for the insights and assistance of Dr. Marty Wazak of the NASA Langley Research Center, Dr. Jared Grauer and Dr. Aimy Wissa formerly of the University of Maryland Morpheus Laboratory, and the authorship and editing assistance of Ms. Debbie McFee. Finally, we are also grateful for the support of the NASA Langley Research Center, the National Institute of Aerospace, the University of Maryland, the Air Force Office of Scientific Research, and the Morpheus Laboratory.

LIST OF FIGURES

Figure 1.1	Flight capability of flapping wing flyers.	2
Figure 1.2	Mission applications for unmanned air vehicles. (A) Civilian and (B) Military.	3
Figure 1.3	(A) Natural flyers demonstrate glide, hover, dive, and perch. (B) Example of maneuverability performance of natural flyers.	4
Figure 1.4	Concept vehicle exhibit adjustable stiffness across the wing.	5
Figure 1.5	Fuselage body accelerations in state-of-the-art flapping wing robotic.	6
Figure 1.6	Position state of fuselage of ornithopter test platform.	7
Figure 1.7	Lift and thrust generating mechanisms of flapping wing flight (A) downstroke and (B) upstroke.	9
Figure 1.8	Wing tip path as indicator or dominant wing motion shown on natural flyers (A) albatross in fast-forward flight mode, (B) pigeon in slow flight mode, (C) horseshoe bat in fast-forward flight mode, and (D) horseshoe bat in slow flight mode.	11
Figure 1.9	Passive wing morphing. (A) Bioinspired test platform wing morphing though thrust flap region and (B) radical shape morphing wing.	16
Figure 1.10	Active wing morphing bioinspired test platform. (A) Ornithopter platform. (B) Extended half wingspan versus retracted half wingspan.	16
Figure 1.11	First ornithopter vehicle designs. (A) DaVincis 1490 and (B) Lilienthal 1894.	17
Figure 2.1	Ornithopter schematic: rigid multi-body system.	24
Figure 2.2	Ornithopter schematic: linear elastic multi-body systems.	24
Figure 2.3	Ornithopter schematic: nonlinear elastic multi-body systems.	25
Figure 2.4	SimXpert: flexible multi-body dynamics model implementation.	31
Figure 2.5	SimXpert: Image of fully integrated flexible multi-body dynamics model of ornithopter.	32
Figure 2.6	Ornithopter vehicle dynamics models. (A) Single rigid body, (B) rigid multi-body dynamics, and (C) flexible multi-body dynamics model.	33
Figure 2.7	Quasi-steady blade element model.	38
Figure 2.8	CFD analysis showing (A) vorticity contours on the flexible deformed wing and (B) deformed grid at 50% span location during upstroke of the wing at start of upstroke.	42
Figure 2.9	Rigid-body ornithopter fit in flight test data for aerodynamic model structure determination.	43
Figure 2.10	Modeling methodology: workflow stage 1.	44
Figure 2.11	Modeling methodology: workflow step 2.	45

Figure 2.12 Modeling methodology: workflow step 3. 45
Figure 2.13 Modeling methodology: workflow stage 4. 45
Figure 2.14 Modeling methodology: workflow stage 5. 46
Figure 2.15 Modeling methodology: workflow stage 6. 46
Figure 2.16 Modeling methodology: workflow stage 7. 46
Figure 3.1 Primary test platform Morpheus Lab custom-build test ornithopter (ML101). 51
Figure 3.2 Secondary test platform modified ornithopter Shawn Kinkade (MSK004). 52
Figure 3.3 Wing structure bioinspired ornithopter test platform (ML101). 52
Figure 3.4 Design feature wing spars (ML101); (A) leading edge spar, (B) diagonal spar, (C) finger spar (ML101). 52
Figure 3.5 Schematic and nomenclature stiffening carbon fiber spar configuration. 52
Figure 3.6 A bat (*Cynopterus brachyotis*) in flight. 53
Figure 3.7 Lower wing surface of a natural avian flyer photo. 54
Figure 3.8 Design feature tail (ML101). 54
Figure 3.9 Design feature fuselage (ML101). 54
Figure 3.10 Definition flapping angle and upstroke and downstroke on shoulder joint/bar linkage (ML101). 56
Figure 3.11 Upstroke and downstroke sequence ornithopter (ML101) scale. 56
Figure 3.12 MSK004 wings with tracking markers. 57
Figure 3.13 MSK004 and Vicon camera system for the measurement of ornithopter wing and configuration motions. 58
Figure 3.14 Harmon aerodynamic model results and bench test results measured integrated forces ML101 at 5 Hz flapping frequency, over a flapping cycle t/T, (A) vertical propulsive force (VPF), (B) horizontal propulsive force (HPF), (C) normalized FA. 59
Figure 3.15 Harmon bench test results measured integrated forces ML101 at 5 Hz flapping frequency, over a flapping cycle t/T, (A) vertical propulsive force (VPF), (B) horizontal propulsive force (HPF) versus normalized flapping angle (FA). 59
Figure 3.16 Harmon aerodynamic model results and bench test results measured integrated forces ML101 at 6.17 Hz flapping frequency, over a flapping cycle t/T, (A) vertical propulsive force (VPF), (B) horizontal propulsive force (HPF), (C) normalized flapping angle (FA). 60
Figure 3.17 Harmon bench test results measured integrated forces ML101 at 6.17 Hz flapping frequency, over a flapping cycle t/T, (A) vertical propulsive force (VPF), (B) horizontal propulsive force versus normalized flapping angle (FA). 60
Figure 3.18 Results and bench test results measured integrated forces ML101 versus flapping frequency, over a flapping cycle t/T, (A) mean absolute value vertical propulsive force (MAVPF), (B) mean horizontal propulsive force (MHPF) versus flapping angle. 61

Figure 3.19 Vicon Vision wing kinematic sequence—three position state tracking markers $i=1-110$, (A) isometric view, (B) side view in a fuselage fixed reference frame CB0. 62

Figure 3.20 Vicon camera system for the measurement of ornithopter wing and configuration motions. 62

Figure 3.21 Aerodynamic forces on the wings—results system ID model by Grauer. *FA*, flapping angle; *HPF*, horizontal propulsive force; *VPF*, vertical propulsive force. 63

Figure 3.22 Asymmetric retroreflective markers on the wing (seen as white dots) and wing tip location (marked in red). 64

Figure 3.23 Retroreflective marker position on the wing measured with Vicon Vision system markers $i=1-53$. Markers $i=1-5$ were mounted on the fuselage. 65

Figure 3.24 Image small Unmanned Aerial System flight test facility—Vicon Vision system cameras. 65

Figure 3.25 Test setup schematic—test chamber dimensions: $70'(W)\times 35'(D) \times 35'(H)$. 66

Figure 3.26 Schematic ornithopter and reference frames used for the processing of free-flight test data. 67

Figure 3.27 Data set marker position during free flight in the inertia reference frame CI0. 68

Figure 3.28 Data set marker position $i=1-53$ during free flight in the inertia reference frame CI0—view ZI0/ZI0. 68

Figure 3.29 Data set marker position $i=1-53$ during free flight in the inertia reference frame CI0—view YI0/XI0. 69

Figure 3.30 Position states all tracking markers on the wing/volume shows test data used for model development—3.5 flapping cycles. 69

Figure 3.31 Single position state on ornithopter shows all tracking markers $i = 1$ to 53 in the fuselage fixed reference frame CB0. 70

Figure 3.32 Vertical propulsive forces acting on the fuselage center of mass—obtained from experiment. *VF*, vertical force; *HF*, horizontal force; FA, flapping angle. 70

Figure 3.33 Vacuum chamber at NASA Langley Research Center. (Left) Test platform mounted on six degrees of freedom load cell, (Right) vacuum chamber. 71

Figure 3.34 Results of vacuum chamber test ML101 test platform versus flapping frequency, (A) inertial horizontal propulsive force IHPF (B) inertial horizontal propulsive force (C) inertial pitching moment (IPM) versus magnitude. 72

Figure 4.1 Schematic five-body dynamics system. 80

Figure 4.2 Schematic inertial and fuselage body fixed reference frames. 81

Figure 4.3 Schematic wing fixed reference frames. 81

Figure 4.4 Articulated multi-body system representation of the ornithopter. 82

Figure 4.5 Position vector of rigid multi-body model of ornithopter. 89
Figure 4.6 Position vector of flexible multi-body model of ornithopter. 90
Figure 4.7 Schematic notation definition for generalized coordinates of the flexible multi-body system. 92
Figure 4.8 Position vector of flexible multi-body model of ornithopter, luff region [blue (darker gray in print versions) and red (light gray in print versions)], thrust flap region [orange (dark gray in print versions) and red (lighter gray in print versions)]. 94
Figure 4.9 Schematic finite element model wing structure carbon fiber spars and wing. 103
Figure 4.10 Thrust flap region mesh nodes = 352, elements = 860. 105
Figure 4.11 Thrust flap region mode shapes contour plot. 105
Figure 4.12 Model image flexible wing component connections. 106
Figure 5.1 Blade element grid schematic aero-model A. 111
Figure 5.2 (A) Blade element grid schematic aero-model B. (B) Blade element (BE) grid schematic aero-model C. 111
Figure 5.3 Variables for calculation of aerodynamic forces. 115
Figure 5.4 ML101 ornithopter configuration blade element (BE) selection. 119
Figure 5.5 Blade element (BE) selection thrust flap region ML101 ornithopter configuration. 119
Figure 5.6 Aerodynamic loads model—workflow. FMBD, flexible multi-body dynamics. 120
Figure 5.7 Aerodynamic loads model—workflow Aero Load Initialized Experimental Coupled. *FMBD*, flexible multi-body dynamics. 122
Figure 6.1 Isometric view of the position states path of all Vicon markers over one flapping cycle at 6.06 Hz. 129
Figure 6.2 Side view of the position states path of all Vicon markers over one flapping cycle at 6.06 Hz. 130
Figure 6.3 Top view of the position states path of all Vicon markers over one flapping cycle at 6.06 Hz. 130
Figure 6.4 Top view of the ornithopter markers position state 0 deflection plane. 131
Figure 6.5 Top view of the *Y* coordinate flexibility in the wing fixed reference plane CW. 132
Figure 6.6 Top view of the *X* coordinate flexibility in the wing fixed reference plane CW. 132
Figure 6.7 Bioinspired ornithopter test platform in free flight: three experimental orientations during a wing beat: viewed in the fuselage body fixed reference frame CB0. 133
Figure 6.8 Wing fixed reference frame CW. 134
Figure 6.9 Bioinspired ornithopter test platform wing in free flight: maximal elastic deflections of the wing in the wing fixed reference frame during a wing beat; viewed in the fuselage fixed reference frame CB0. 135

Figure 6.10 Bioinspired ornithopter test platform wing in free flight: maximal occurring elastic deformation in the thrust flap region of the wing in the wing fixed reference frame CW. 135

Figure 6.11 Bioinspired ornithopter test platform ML101 wing in free flight: maximal occurring elastic deflections luff region in the wing fixed reference frame CW. 136

Figure 6.12 Bioinspired ornithopter test platform wing in free flight: maximal occurring elastic deflections of the wing fixed reference frame during a wing beat; viewed in the fuselage reference frame CB0. 136

Figure 6.13 Schematic location thrust flap region reference frame CBT. 137

Figure 6.14 Bioinspired ornithopter test platform wing in free flight: *X*-axis of modeled thrust flap region reference frame in 34 position states viewed in the fuselage fixed reference frame CB0. 137

Figure 6.15 Bioinspired ornithopter test platform wing in free-flight marker location: XBT-axis location on experimental test platform resulting in minimal deformation of a reference YBT/ZBT reference plane using a 1 degree of freedom flapping motion. 138

Figure 6.16 Flapping angle zeta and beta location on shown in experimental wing test data (maximal deformation of CW,0 reference plane, and XBT position states). 139

Figure 6.17 Modeled flapping angle (FA) time history based on flight test data versus FA in flight test data E-2. 139

Figure 6.18 Experimental thrust flap reference motion in 34 position states in the fuselage fixed reference plane CB0. 140

Figure 6.19 Modeled thrust flap reference motion in 34 position states in the fuselage fixed reference plane CB0 versus wing at zero deformation (PS0). 141

Figure 6.20 Experimental elastic deflections in the fuselage fixed reference frame CB0 during a wing beat. 142

Figure 6.21 Experimental elastic deformation in the wing fixed reference frame CW during a wing beat. 142

Figure 6.22 Model image SimXpert—ML101 five-body flexible multi-body dynamics model (fuselage reference frame CB is highlighted). 143

Figure 6.23 Verification bench test—model 3 A—scale ML101 (6.17 Hz). *FMBD*, flexible multi-body dynamics; *HFI*, horizontal force inertia; *HPF*, horizontal propulsive force; *VPF*, vertical propulsive force. 145

Figure 6.24 Wing tip location—viewed at one experimental position state (mid-downstroke). 146

Figure 6.25 Results wing kinematics model 5 experiment: flight test: (left) back view and (right) side view. 147

Figure 6.26 Propulsive forces acting on the fuselage center of mass—obtained from experiment (E-2) and five-body vehicle dynamics model using aerodynamic model C. 148

Figure 6.27 Simulated inertia forces on fuselage center of mass CB in free flight (inertial vertical propulsive force)—upstroke–downstroke transition. 149

Figure 6.28 Results inertia forces (E3) with error bars versus simulation results with STD, (A) inertial horizontal propulsive force (IHPF), (B) inertial vertical propulsive force (IVPF), (C) inertial pitching moment (IPM) versus magnitude. 150

Figure 6.29 Results: body force magnitude system ID model (MSK004 E1-I)—body force magnitude—aero-model C. 152

LIST OF TABLES

Table 1.1 Properties of natural flyers in comparison to well-known aircraft 15

Table 2.1 Finite element method (FEM) and multi-body dynamics (MBD) university codes and commercial packages 29

Table 3.1 Material and dimensions of ornithopter structures 55

Table 3.2 Aircraft specifications 55

Table 4.1 Material properties and dimensions of ornithopter structures used for the model 104

Table 5.1 Aerodynamic model versions overview 110

Table 5.2 Blade element selection ML101 ornithopter configuration 118

LIST OF NOMENCLATURE

ROMAN SYMBOLS

a	Location of airfoil pitching axis from 0.5 chord point
b	Blade element
c	Reference chord
dc	Blade element chord length
dr	Blade element with span direction
f_n	Force vector
f_g	Generalized gravitational force
f_e	Externally applied forces
f_v	Coriolis and centrifugal forces
f_c	Constraint forces
i	Number of bodies
u_B	Interface DOF
u_I	Interior DOF
k	Denotes rigid link
m	Number of rigid links
n	Number of blade elements
nm	Number of nodes a force is applied on Bodies III (L/R)
nl	Number of nodes a force is applied on Bodies II (L/R)
m_{pp}	Node's mass
$\overrightarrow{p}_{B0A^{n}}$	Position vector from the origin of the local body reference frame CBW (B0) of Bodies II (L/R) to the nodal point's nl of force application
$\overrightarrow{p}_{TA^{nm}}$	Position vector from the origin of the local body reference frame CBT (T) to the nodal point's nm of force application on Bodies III (L/R)
$\mathbf{q}$	Vector of modal coordinates
$q*$	Orthogonalized Craig–Bampton modes
q_c	Modal coordinates of the constraint modes
q_n	Modal coordinates of the flexible-boundary modes
$\mathbf{q}_i$	Modal coordinates
q_f	Elastic coordinates can be introduced using component mode methods
$\overrightarrow{r}_{B0P}$	Position vector of the unreformed position of point P with respect to the local fuselage body reference frame of Body B0
$\overrightarrow{r}_{BB0}$	Position vector from the fuselage fixed reference frame CB to the fuselage fixed reference frame CB0
$\overrightarrow{r}_{IP}$	Position vector origin of the inertial (ground) reference frame CI to the unreformed location of marker P on the rigid wing body
$\overrightarrow{r}_{IB}$	Position vector from the origin of the inertial (ground) reference frame to the origin of the local fuselage body fixed reference frame CB

$\vec{r}_{PP'}$	Translational deformation vector of point P, the position vector from its undeformed to its deformed position P'
$\vec{r}_{B0TR}$	Position vector from the fuselage fixed reference frame B0 to the wing fixed reference frame TR
$\vec{r}_{TRP}$	Position vector of the undeformed position of point P with respect to the wing fixed reference frame of the wing Body II
$s(t)$	Shape function
u_B	Interface DOF
u_I	Interior DOF
$v_{k,}$	Linear velocity of center of mass of link k
$v(\xi_{IB(E)})$	Experimental free flight velocity of the vehicle fuselage fixed reference frame B in the inertial reference frame I
v_p	Undeformed location
x	Physical nodal DOF
x, y, z	Cartesian coordinates
x_k	Position coordinates of link k

CAPITAL ROMAN SYMBOLS

A	Transformation matrix from the local body reference frame of B0 (fuselage body fixed) to the inertial reference frame
C	Energy dissipation function
C_d	Drag coefficient
$C(\xi\dot{\xi})$	Coriolis and centrifugal forces in terms of the generalized coordinates
D_{camber}	Chordwise force due to camber
D_f	Chord-friction drag
F_L	Force vector
F_{horiz_nc}	Noncirculatory force horizontal component
F_{vert_nc}	Noncirculatory force vertical component
I	Inertia invariants
J	Jacobian matrix
K	Stiffness matrix
$K(\xi)$	The stiffness matrix in terms of the generalized coordinate
K_{tt}	Translational slice of the time-varying stiffness matrix
K_{oo}	Rotational slice of the time-varying stiffness matrix
K_{mm}	Modal slice of the time-varying stiffness matrix
K_{mt}	Modal and translational cross-terms in the time-varying stiffness matrix
K_{ot}	Rotational and translational cross-terms in the time-varying stiffness matrix
K_{mo}	Modal and rotational cross-terms in the time-varying stiffness matrix
$\hat{K}$	Generalized stiffness matrix

$\widehat{K}_{NN}$	Diagonal stiffness associated with eigenvectors
K_{ff}	Stiffness matrix due to elasticity
L	Marker L
L	Lagrange
M	Mass matrix
M_{ff}	Flexible slice of the time-varying mass matrix
M_{rr}	Rigid-body of the time-varying mass matrix
M_{fr}	Flexible and rigid-body cross-terms in the time-varying mass matrix
M_{rf}	Rigid-body and flexible cross-terms in the time-varying mass matrix
$M(\xi)$	Mass matrix of the generalized coordinate
M_{tt}	Translational slice of the time-varying mass matrix
M_{oo}	Rotational slice of the time-varying mass matrix
M_{mm}	Modal slice of the time-varying mass matrix
M_{mt}	Modal and translational cross-terms in the time-varying mass matrix
M_{ot}	Rotational and translational cross-terms in the time-varying mass matrix
M_{mo}	Modal and rotational cross-terms in the time-varying mass matrix
$\widehat{M}_{NN}$	Diagonal mass matrices associated with eigenvectors
$\widehat{M}$	Generalized mass matrix
ME_L	Moment vector
N	Number of modes
dN	Normal force
N_a	Noncirculatory normal force
$N(\xi)$	Constraint force
N_{nc}	Noncirculatory force
N_c	Circulatory normal force
Q	Generalized forces
$Q(\xi)$	Generalized applied force projected to the generalized coordinate
Q_F	Generalized translational force
Q_{ME}	Generalized moments
Q_M	Generalized modal force
Q_F^I	For the total integrated forces in the inertia reference frame due to applied aerodynamics loads
Q_{FIIIL}^{CB}	Individually generalized translational force on Bodies III on a marker L/node nm
Q_{FIIL}^{CB}	Individually generalized translational force on Bodies II on a marker L/node nl
Q_{FI}^{CB}	Generalized translational force on Body I
R_k	Orientation coordinates of link k
T	Kinetic energy
T_s	Leading edge suction force
V	Potential energy

$V_{0.25c}$	Velocity at the quarter chord location
V_x	The flow speed tangential to the section

GREEK SYMBOLS

α'	Flow's relative angle of attack at the 0.75 chord location
α_0	Zero lift angle of attack
β	Flapping angle beta
$\ddot{\beta}$	Angular acceleration of the flapping wing
γ	Flight path angle of flight velocity with respect to the ground
δ	Angel between flapping axis and flight velocity
ε	Vector of small angles
ζ	Thrust flap angle zeta
ξ	Generalized coordinates
ξ_I	Generalized coordinates of the fuselage Body I
$\xi_{II(L/R)}$	Generalized coordinates of the flexible luff region wing Bodies II (R/L)
$\xi_{III(L/R)}$	Generalized coordinates of the flexible luff region wing Bodies III (R/L)
$\xi_{IB(E)}$	Experimentally generalized coordinates of the fuselage in the inertia reference frame
$\xi_{BBW(E)}$	Experimentally generalized coordinates of the wing bodies in the inertia reference frame in the wing region
$\xi_{BWTW(E)}$	Experimentally generalized coordinates of the wing bodies in the inertia reference frame in the thrust flap region
ξ^i	Generalized coordinate of the flexible body i
ξ_r	Generalized coordinate of a rigid body
η_s	Leading edge suction efficiency
θ	Pitch angle of blade element with respect to flapping axis
$\dot{\theta}$	Pitching velocity of the blade element
$\overline{\theta}$	Total mean pitch angle
λ	Lagrange multipliers
ϕ	Shape vectors (or mode shapes)
$\psi, \theta, \phi.$	Euler angles
ω_k	Angular velocity of center of mass of the rigid link k
$\omega_{p(k)}$	Angular velocity of the parent link
Φ_p	The slice from the modal matrix that corresponds to the translational DOF of node P
Φ_{p*}	The slice from the modal matrix that corresponds to the rotational degrees of freedom (DOFs) of the node P
Φ_p	Participation in the component modes

Φ_{IC}	Physical displacements of the interior DOF in the constraint modes
Φ_{IN}	Physical displacements of the interior DOF in the normal modes
Ψ	Algebraic constraint equation

SUPERSCRIPTS AND SUBSCRIPTS

B	Boundary DOF
C	Constraint DOF
I	Internal DOF
N	Normal DOF
f	Elastic coordinates
m	Modal degrees of freedom
nc	Circulatory force
r	Reference coordinates
o	Rotational degrees of freedom
t	Translational degrees of freedom

ACRONYMS

AEOI	Aeroelastic open interface
AFRL	Air Force Research Laboratory
ANCF	Absolute nodal coordinate formulation
AOA	Angle of attack
BE	Blade element
BEM	Boundary element method
CAD	Computer-aided design
CFD	Computational fluid dynamics
DAE	Differential and algebraic equations
DOF	Degrees of freedom
E1	Experiment 1
E1-I	Experiment 1—inertial
E2	Experiment 2
E3	Experiment 3
EMBS	Elastic multi-body system
FE	Finite elements
FEM	Finite element method
FEIM	Finite element incremental method
FFR	Floating frame of reference formulation
FSI	Fluid structure interaction

FMBD	Flexible multi-body dynamics
FMBSDA	Flexible multi-body structural dynamic model considering aeroelasticity
FSM	Finite segment method
GUI	Graphical User Interface
HF	Horizontal force
HIF	Horizontal inertia force
HPF	Horizontal propulsive force
LEV	Leading edge vortecies
LMBS	Linear multi-body system
LRVF	Large rotation vector formulations
LTED	Linear theory of elastodynamics
LTI	Linear time invariant
NLTP	Nonlinear time periodic
MAVs	Microaerial vehicles
MBD	Multi-body dynamics
MP	Multi-physics
ODEs	Ordinary differential equations
PDEs	Partial differential equations
RMBD	Rigid multi-body dynamics
SUAVs	Small unmanned aerial vehicles
TVL	Time-variant loads
TVSD	Time-variant and spatially distributed
TiVs	Tip vortecies
VF	Vertical force
VLMs	Vortex lattice methods
VIF	Vertical inertia force
VPF	Vertical propulsive force
WPAFB	Wright-Patterson Air Force Base

ACKNOWLEDGMENTS

The authors wish to thank NASA Langley Research Center (LaRC), the University of Maryland, and the National Institute of Aerospace (NIA) for their continued support. Funding for this work was supplied by LaRC.

SUMMARY

Small aerial robots are needed to conduct a variety of civilian and military missions creating the need for researchers to understand the physics of avian-based flapping wing vehicles, known as ornithopters. Efforts to model the flight physics of these vehicles are complicated by a number of factors, including nonlinear elastic effects, multibody characteristics, unsteady aerodynamics, and the strong coupling between fluid and structural dynamics. Experimental verification is crucial to achieve accurate simulation capabilities.

A multidisciplinary approach to modeling requires the use of tools representing individual disciplines, which must be combined to form a comprehensive model. In the framework of this work a *five-body flexible-vehicle dynamics model* and a novel experimental verification methodology are presented. For the model development and verification of the modeling assumptions, a data set providing refined wing kinematics of a test ornithopter test platform in free flight was used. Wing kinematics for the verification was obtained using a Vicon motion capture system.

Lagrange equations of motion in terms of a generalized coordinate vector of the rigid and flexible bodies were formulated to model the flexible multibody system. Model development and verification results are presented.

The "luff region" and "thrust flap region" of the wing were modeled as flexible bodies. A floating reference frame formulation was used for the ornithopter. Flexible body constraints and modes were implemented using the Craig—Bampton method, which incorporates a semiphysical subspace method. A quasi-steady aerodynamic model using Blade Element Theory was correlated and verified for the problem using the experimental wing kinematics. The aerodynamic model was then formulated in terms of generalized coordinates of the five-body flexible multibody system and was used in the resulting model to account for aeroelasticity. Modeling assumptions were verified, and simulation results were compared with experimental free flight test data.

CHAPTER ONE

Bioinspired Flight Robotics Systems

1.1 INTRODUCTION OF THIS BODY OF WORK

Flapping wing aerial vehicles have the potential to be utilized for search and rescue missions, environmental surveillance, and to aid first responders. The body of work presented here leads to a flexible multi-body dynamic representation of a flapping wing ornithopter considering aeroelasticity. This body of work brings advances in the understanding of flapping wing flight physics and dynamics that ultimately leads to an improvement in the performance of such flight vehicles, and thus allowing them to reach their high performance potential. To accomplish the complex problem of improving ornithopter flight dynamics, a novel approach to flexible multi-body dynamics and aeroelasticity modeling is presented in conjunction with experimental data. Principles of flapping wing flight strategies and the simulation method developed herein might also be employed in the future to enhance the performance of other aircraft to lower energy consumption and lead to more eco-friendly flight.

1.2 THE BACKGROUND OF FLAPPING WING FLIGHT TECHNOLOGY

1.2.1 Aerial Vehicles and Natural Flapping Wing Flyers

Small insect scale flyers, microaerial vehicles (MAVs), small unmanned air vehicles (SUAVs), and avian scale flyers have generated enormous interest in recent years due to a variety of potential applications in the civilian and military sector [1,2].

Research and Development efforts for flapping wing flight have mainly focused on MAVs due to the advantageous of aerodynamic efficiency associated with flapping wing vehicles operating in this flight regime [3—5]. Simple wing gates and dual flight configurations of hover and straight-and-level flight dominate this regime for flapping wing MAVs. Flapping wing flight on the avian scale is not entirely understood and has great performance potential due to its multi-flight configuration capability. MAV's

Modern Flexible Multi-Body Dynamics Modeling Methodology for Flapping Wing Vehicles
ISBN: 978-0-12-814136-6
http://dx.doi.org/10.1016/B978-0-12-814136-6.00001-9

wing gates and associated kinematics are complicated and occur in a transition Reynolds number flight regime. Here the aerodynamic advantages associated with flapping wings can be utilized as well to allow for efficient gliding flight. The associated wing gates (wing motion profiles), aerodynamics, and dynamics are not yet entirely understood [6] here or in SUAVs, which are of slightly larger scale. SUAVs lie on the opposite end of the flight spectrum from the conventional systems currently in use today. Flapping wing aerial vehicles fall in the SUAVs category. These vehicles combine the ability to hover like rotary-wing aircraft (as demonstrated by AeroVironment's hummingbird), while also allowing for gliding flight, much like fixed-wing aircraft [7].

Within the domain of SUAVs, the aerodynamics associated with flapping wing platforms demonstrate optimal properties characterized by small vehicle size and low Reynolds number flight [6,8]. Flapping wing flight vehicles have the capability to combine the three sides of the performance triangle: (1) ideal aerodynamic performance at a low Reynolds number flight regime, (2) agility and maneuverability, and (3) mission adaptability in one vehicle, as illustrated in Fig. 1.1. Consequently, they fill a niche in the design space of SUAV's [9].

Flapping wing vehicles have the ability to dive and perch, are highly maneuverable and agile, and have improved safety and reduced noise

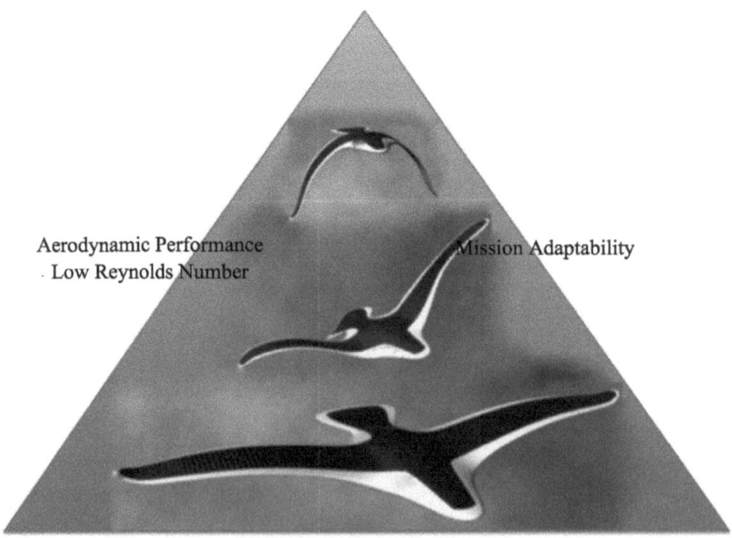

Figure 1.1 Flight capability of flapping wing flyers [10].

emissions when compared to rotary-wing vehicles. Additionally, flapping wing vehicles have visual properties, which make them ideal for contextual camouflage. These qualities and flight dynamics make them suitable, sustainable, and ideal for a variety of civilian and military mission profiles. These vehicles have the capability to perform specialized tasks such as gathering environmental information; atmospheric data collection; aerial surveillance; homeland security; and supply, search, and rescue missions. They can also aid policemen and firefighters to perform dull, dirty, and dangerous jobs as well as act as supporting team members for surveillance tasks (Fig. 1.2).

The flapping wing flight strategy for lift and thrust generation is key to enabling technology in the varied multi-mission capability of SUAV ornithopters. At relatively light wing loadings, flapping wing SUAVs are more aerodynamically efficient than conventional fixed-wing or rotary-wing vehicles [4,5]. As the vehicle size decreases, viscous effects become more pronounced and fixed-wing vehicles can suffer from decreased lift to drag ratios and decreased flight performance [3,4]. For rotary-wing aircraft, viscous effects reduce the aerodynamic efficiency of the vehicle [3]. Research by Wang et al. demonstrated that when optimal wing motions are applied in low Reynolds number flight, flapping wing technologies can save up to 27% of the aerodynamic power required by fixed- and rotary-wing vehicles. This indicates that the aerodynamic power needed to support a specified weight is lower when using optimal flapping wing motions [8]. In the design space of SUAVs, simple single flight mode mission profiles can be satisfied with either rotary- or fixed-wing vehicles. However, novel vehicle designs are needed to enable vehicles to fly agilely in obstacle-filled and complex environments as well as efficiently in straight-and-level flight [9].

Figure 1.2 Mission applications for unmanned air vehicles. (A) Civilian [12] and (B) Military [11].

(A)

(B)

Figure 1.3 (A) Natural flyers demonstrate glide, hover, dive, and perch [18]. (B) Example of maneuverability performance of natural flyers [18].

Mechanical flapping wing aerial flyers demonstrate varying degrees of ability in terms of accomplished flight performance [2,13,14]. The University of Maryland's Morpheus Lab ornithopters are among the most successful flapping wing research flight platforms [15–17]. However, mimicking the flight of natural flyers is still a significant challenge to flapping wing designers and researchers due to the lack of a full understanding of the complex, nonlinear, and time-varying dynamics involved [9].

The maneuverability, agility, and duration of flight when exploiting atmospheric thermals and natural wing gate dynamics are still not achieved in mechanical flyers. Some desired capabilities are glide, hover, dive, and perch, such as those exhibited by birds. Fig. 1.3 shows maneuverability performance.

1.3 A MODEL OF AN ORNITHOPTER FOR PERFORMANCE OPTIMIZATION

To reach the performance capabilities of natural flyers, a flexible multi-body dynamics model of an ornithopter considering aeroelasticity was developed in this body of work. The focus of the modeling effort was on the capability of improvement of flight characteristics in straight-and-level flight. In using this model, it was desirable to reduce body accelerations and forces of an ornithopter vehicle, as well as to improve the aerodynamic performance, and enhance flight kinematics and forces, which are the design optimization objectives. Achieving these objectives will result in the establishment of new requirements and model capabilities.

The most influential factors identified in the open literature and empirical observations that influence performance objectives and their improvements (defined later in this chapter) are the wing geometry, wing configuration, the wing motion profile, and the wing flexibility [2,6,8,19,20].

Flapping wings with suitable wing kinematics, wing shapes, and flexible structures can enhance lift as well as thrust by exploiting large-scale vortical flow structures under various conditions [6].

The relevant model input variables are illustrated in Fig. 1.4 on the anatomy of a concept vehicle, which accentuates mentioned influence parameters like in natural flapping wing flyer.

In summary, wing geometry, wing motion profile, and wing flexibility are variable and become input parameters in the model to be used to optimize performance of these vehicles in straight-and-level flight. These parameters are discussed in the context of the aerodynamics of flapping wing and bioinspired flapping wing flight. Desired improvements in dynamic characteristics of flapping wing flight vehicles are evaluated first, followed by a brief explanation of aerodynamic mechanisms of flapping wing flight.

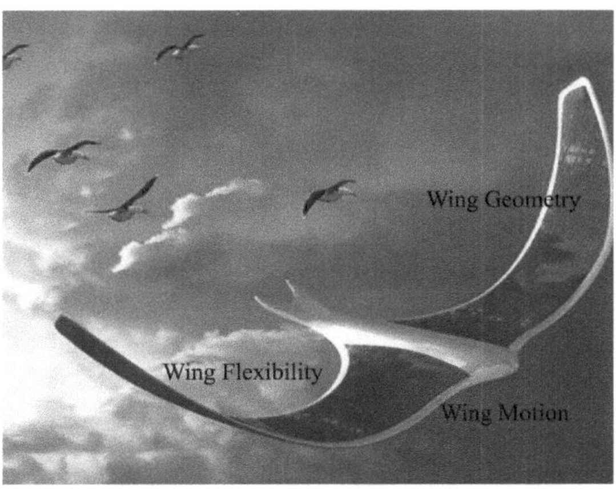

Figure 1.4 Concept vehicle exhibit adjustable stiffness across the wing [21].

1.3.1 Desired Improvements in Flight Platforms

It is of interest to smooth the vehicle body dynamics, i.e., reduce the pitch and heave motions, reduce associated forces in vehicle platforms, and improve the aerodynamic efficiency.

Reducing body forces and accelerations in ornithopters are desired to build lighter vehicle platforms and improve imaging capabilities. In current state-of-the-art ornithopter platforms, body accelerations up to 4 g's are found during straight-and-level flight. Fig. 1.5 illustrates body accelerations on a state-of-the-art experimental ornithopter. Most desired mission scenarios for ornithopter platforms include aerial surveillance. An enhanced imaging capability can be achieved by reducing the body accelerations and forces. Here lighter, less robust onboard camera systems will be feasible. Lesser postprocessing and reduced onboard imaging postprocessing equipment is required. Currently this is achieved through robust and highly specialized systems. With an improvement of body forces and accelerations, payloads can be more fragile. State-of-the-art ornithopter flight platforms require customized and robust onboard avionics [17]. With the improvement of body dynamics, lighter and less robust and customized onboard avionics can be used, leading to a desired reduction in vehicle weight. This in turn leads to an improvement of range and endurance or an increase in payload capabilities.

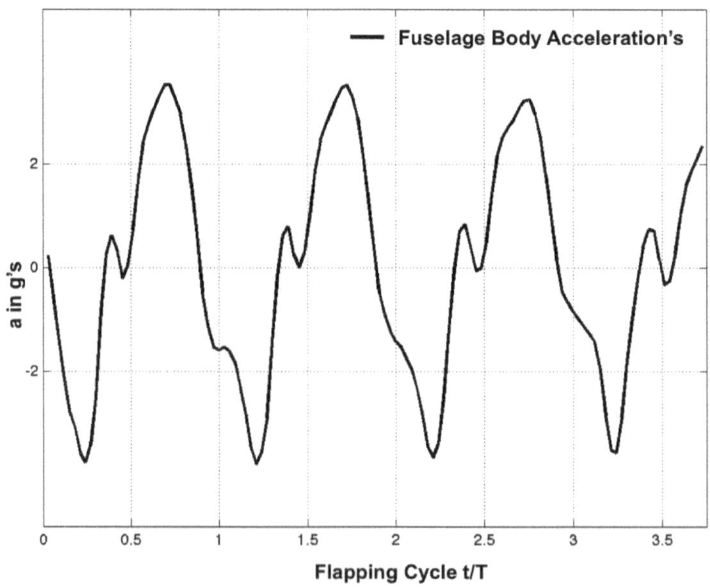

Figure 1.5 Fuselage body accelerations in state-of-the-art flapping wing robotic.

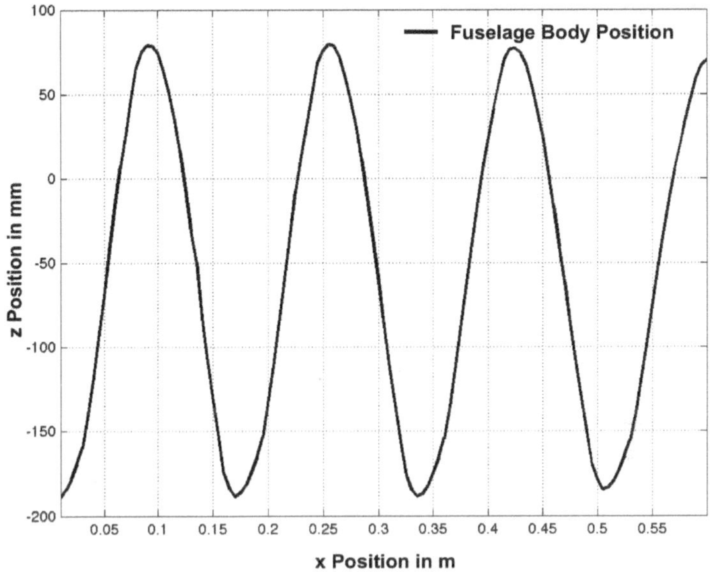

Figure 1.6 Position state of fuselage of ornithopter test platform.

In summary, reduced body dynamics will allow less customized on board cameras, processing, vehicle control, and avionics packages to be used for vehicle designs. These factors ultimately drive the cost down and make flapping wing more feasible for commercialization.

Reduction of the oscillations on the fuselage kinematics allows one to observe an oscillation in the position state of the fuselage of a state-of-the art ornithopter platform in Fig. 1.6. This is associated with the reduction of variations in the lift and thrust profile, which is consequently desired.

Natural flyers exhibit reduced variations in their lift and thrust profile as compared with traditional vehicle platforms. A reduction in such variations leads to better contextual camouflage as well as improved dynamics and consequently simplified flight control algorithms. It is also of interest to increase the aerodynamic efficiency of the flight vehicles and therefore improve the specific power required to generate lift and thrust. Significant characteristics that influence aerodynamic performance and basic aerodynamics in straight-and-level flight are discussed later.

1.3.2 Background and Flapping Wing Flight Aerodynamics

The aerodynamics of flapping wing flight is governed by steady and unsteady lift and thrust generating mechanisms [22]. The unsteady effects and circulatory forces that influence the flow structure for flapping wing flight are largely responsible for aerodynamic force generation. Unsteady flow mechanisms and their contribution in the flapping wing flight performance are not entirely understood [20,22,23]. Basic effects to form advantageous flow structures, their interactions, and related open issues on a two-dimensional and three-dimensional wing as well as a membrane wing structure are discussed. Outcomes such as apparent mass effects, induced flows, wake capture, and dynamics stall contribute to the flight dynamics of avian scale flapping wing flyers. Flow structures such as the leading edge vortecies (LEVs) and tip vortecies (TiVs) behaviors influence these outcomes. Vortex behavior is largely guided by wing kinematics and wing design, which is a variable in an ornithopter model for its design optimization. To categorize the influence of magnitude of mechanisms forming the aerodynamic flow structure in flapping wing flight, the Reynolds number is used as similarity parameter. Reynolds number dimension, for example, changes the LEV formation and among other features accounts indirectly for wing size and flapping frequency, which is significant for this work. Flapping flight TiVs can interact with the LEV to enhance lift without increasing the power requirements [6,24]. It is established that a change in Reynolds number, for example, caused by wing sizing and flapping frequency leads to a change in the LEV. The spanwise flow structure is also influenced by a change in the Reynolds number. LEV, TiVs, and the spanwise flow structure impact the aerodynamic force generation [6,24]. In flapping wing flight, unsteady effects such as the TiVs and LEV interaction can be exploited to improve the specific lift, while in classical steady aerodynamics, TiVs take away from the energy required for lift production [6]. More details about flapping wing flight effects can also be found in the Reference section of this document [20,22,24,25], which provides a comprehensive review.

The general aerodynamic force generation of flapping wing flight through wing beating is illustrated on an airfoil in Fig. 1.7 and described below. In principle, a horizontal propulsive force is achieved though a forward pointing force vector during a negative angle of attack (AoA) while in the downstroke (Fig. 1.7A). During the upstroke of the wing the generated force vector points backwards in the opposite direction of the desired flight velocity (Fig. 1.7B) and produces a negative thrust component at the positive AoA. It is apparent that the time history of the wing motion, hence kinematics, plays an important role to achieve a positive average horizontal propulsive force during a wing beat for forward flight.

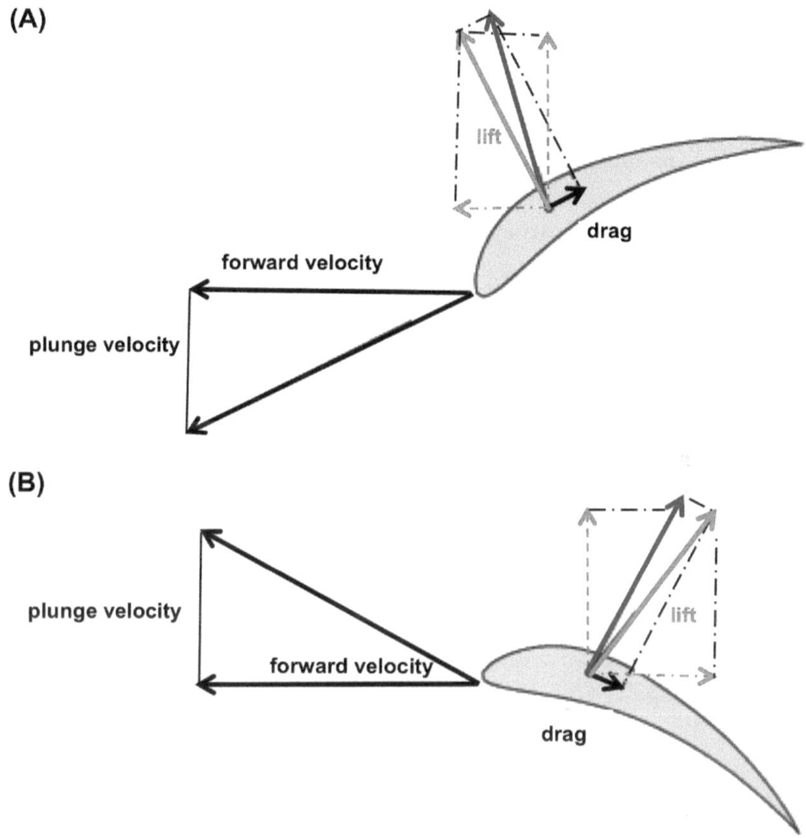

Figure 1.7 Lift and thrust generating mechanisms of flapping wing flight (A) downstroke and (B) upstroke [9].

For a fixed airfoil shape (two-dimensional flapping and plunging), the aerodynamic force generation depends on the time history of AoA during a wing beat and on the shape of the airfoil similar to classical stationary wing theories (Fig. 1.7). For a membrane-like wing structure, the magnitude and orientation of the lift, thrust force, and drag component at an instance of time depend on the form of the camber and the AoA. Flapping cycle camber forming as well as the time history of AoA during a wing beat plays an important role for the resulting flow structure and therefore aerodynamic force generation. The two time-variant geometrical properties (camber form, AoA) indicate the many possible performance combinations employed in a flexible wing structure.

The possibilities of combinations for aerodynamic force generating effects are further increased when extended to a three-dimensional wing.

Here the shape can also differ in the spanwise direction and generates new time-varying lifting and control surfaces. Spanwise flexibility in forward flight creates a varied shape along the membrane wingspan resulting in varied phase shift and effective AoA distribution from the wing root to the wing tip [6]. Such a flapping wing design can also include passive or active joint motions in its three-dimensional wing topology, hence the aerodynamic performance possibilities are even further extended.

1.3.2.1 Motion Profile and Wing Gates

Inherent lift and thrust depend on the geometrical properties and kinematics of the flapping wing. These are primarily governed by wing flexibility and therefore elastic structural properties of the wing as well as wing motion profiles and the wing shape. Flexibility and motion profiles determine the camber forming, which occurs during a wing stroke and the cord and spanwise geometrical properties of the lifting surface during a wing stroke. Wing motions are the primary driver of the resulting unsteady aerodynamics associated with flapping wing flight [6].

Flapping wing kinematics fall into the biokinetics category of flying and swimming [20]. For example, certain fish also operate in a comparable Reynolds number regime and similarly exhibit unsteady motion. A constant or a small variation in thrust profile is accomplished with the unsteady motions, i.e., a fish does not speed up and slow down due to its body motion but rather swims at a constant speed. This same effect is desired in flapping wing vehicles to reduce oscillations in the lift and thrust profile as well as pitch and heave motion.

Successful flyers exhibit wing motion profiles or so-called wing gates. During such gaits, the wing tip path serves as an indicator of the dominant wing displacements and determines wing gates in the field of flapping wing research. Different kinds of wing motions are found in different species as well as during different flight modes such as hover, straight-and-level flight, take off, and landing [22]. Wing tip paths exhibited by natural flyers during one wing beat in straight-and-level flight can be observed and are highlighted in Fig. 1.8.

The velocity and accelerations needed to follow these trajectories as well as the wing path itself are crucial parameters in the aerodynamic performance of flapping wing flyers. Knowledge of such optimal velocities and accelerations is desired for flapping wing vehicle platforms to improve aerodynamic efficiency for the flight mode. Further details about wing gates can be found in Refs. [20,22].

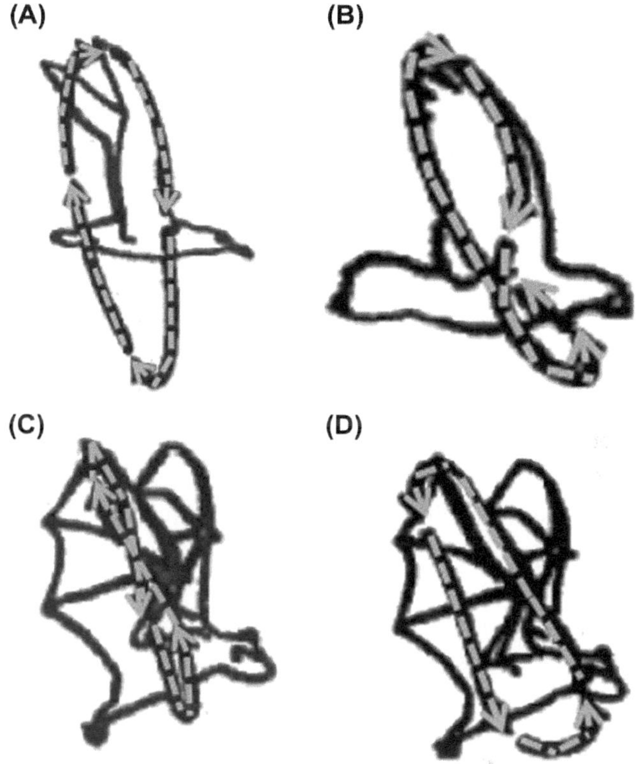

Figure 1.8 Wing tip path as indicator or dominant wing motion shown on natural flyers (A) albatross in fast-forward flight mode, (B) pigeon in slow flight mode, (C) horseshoe bat in fast-forward flight mode, and (D) horseshoe bat in slow flight mode [24].

Based on Wang's research on two-dimensional flapping motion, it can be expected that performing various nonsinusoidal wing motion profiles will lead to an enhanced duration of flight through a lower power requirement for lift and thrust production [8].

A variable wing motion capability needs to be included in the modeling and simulation of such platforms to enhance performance in straight-and-level flight. From the work presented herein, the resulting design model is capable of varying flapping motion profiles and can predict the resulting aerodynamic force generation on the wing. In return the model developed here can be employed for an optimization of aerodynamic efficiency of the overall three-dimensional wing by varying the actuating wing kinematics.

1.3.2.2 Wing Flexibility

It is widely understood that natural flyers such as insects, birds, and bats have flexible wings to adapt to the surrounding flow environment [24]. Natural flyers, such as bats, have deforming bones and anisotropic wing membranes, which have substantial variations of (adjustable) stiffness across the wings (see Fig. 1.4) [2,19,26].

Also it has been shown that the flexible wing has a higher thrust-to-power ratio than its rigid counterpart [6]. Previously, Gopalakrishanan analyzed the effects of elastic cambering of a rectangular membrane wing on the aerodynamics of flapping wing vehicles in forward flight [27]. Different membrane prestresses were investigated to give a desired camber in response to the aerodynamic loading. Results showed that the camber introduced by the wing flexibility increased the thrust and lift production considerably. These results show that the membrane wing outperforms the rigid wing in terms of lift, thrust, and propulsive efficiency for various wing stiffnesses [27].

The increase in force production is primarily attributed to the gliding of the LEV along the camber, which results in a high-pressure difference across the wing surface. Analysis of the flow structures revealed that, for flexible wings, the LEV stayed attached on the top wing surface and covered a major area of the wing, which resulted in high force production. On the other hand, the LEV interacting with the surface of the rigid wing resulted in low force production. In addition, high stiffness in the spanwise direction and low stiffness in the chordwise direction resulted in a uniform camber and high lift and thrust production. A flexible wing membrane also stalls at significantly higher angles of attack compared to a rigid configuration. This has been established for MAV's at a very low Reynolds number and was conclusive for a nonflapping wing configuration [28]. These results when translated to flapping wings identified a need for additional research. Low aspect ratio rigid wings have lower lift curve slopes when achieving the same stall angle of the flexible wing [29]. The flexible properties of the wing can be used for flight quality improvement and can effectively maintain desirable lift characteristics while having improved stall margins [24,30]. Wing forming due to flexibility is significant for its shape manipulation and reconfiguration, and therefore it aids maneuvering and improves agility. Flexible wings form control surfaces and can be used to enhance stability and control [28]. Garzia et al. developed a flapping wing MAV, which improved agility performance due to its wing flexibility [31].

In regards to aeroelasticity, chordwise flexibility in the forward flight can substantially adjust the projected area normal to the flight trajectory via shape deformation, and therefore redistribute lift and thrust [6].

In summary, the wing flexibility is a significant property that influences the aerodynamic force generation and therefore the performance of ornithopter, therefore, has to be included in the model capability. The model developed herein enables an enhancement of the understanding of the influence of variable wing stiffness on the performance.

1.3.2.3 Wing Geometry

The following section reviews the influence and implications of wing geometry in the straight-and-level flight aerodynamic performance of an ornithopter. A numerical investigation by Lentink and Gerritsma concluded that the thin airfoil with aft camber outperformed other airfoils including the more conventional airfoil shapes with thick and blunt leading edges [6,32]. Flow was computed around plunging airfoils at Re of order 10^2 [6,32]. Research by Usherwood and Ellington concluded a minor influence of aspect ratio at AoAs below 50. They performed an experimental investigation using an experimental Hawkmoth. Their test platform had adjusted aspect ratios ranging from 4.53 to 15.84, with corresponding Re of 1.1×10^3 to 2.6×10^4 [6,33]. Green and Smith investigated 3-D effects using particle image velocimetry (PIV) measurements and unsteady pressure distributions on a pitching flat plate in forward flight, at a Reynolds Number between 3.5×10^3 and 4.3×10^4 and aspect ratios of 0.54 and 2.25. They concluded that three-dimensional effects increased with decreasing aspect ratio or with increasing pitching amplitude [34].

An inviscid model by Athshuler illustrated that the aspect ratio for hovering flapping wing vehicles is more significant. An increased aspect ratio and increased wing length and wing area enhances the lift [35,36]. Altshuler et al. also demonstrated for straight and level performance, that lift tends to increase, and lift to drag ratios improve, as wing models approach natural configurations in terms of leading edge sharpness and having a substantial camber. The effect of wing shape was experimentally tested at Reynolds numbers revolving between 5×10^3 and 2×10^4 [6,35].

The airfoil shape effect when comparing the flow field around a pitching and plunging airfoil (SD7003) to a flat plate was investigated by Kang et al. in forward flight. They used (computational fluid dynamics) CFD and PIV measurements to investigate the formation of the flow structure around the wing shapes. It was shown that the sharp leading edge of the flat plate lead to a flow separation at all phases of motion because the flow was not able to turn around as on the airfoil structure with the blunter leading edge. The flow separation induced larger vertical flow structures on the suction side of the flat plate, leading to coverage of an increased area of lower pressure distribution there. The vertical flow structures are responsible for an increased lift generation on the flat plate compared to the SD7003 airfoil shape [6,37].

The implications of this work include the capability of using a variation in wing geometry to enhance forward flight aerodynamic performance and the possibility that advantageous aerodynamic effects could be developed using a variety of wing shapes. Baseline models include an aft camber, a thin airfoil structure, and sharp leading edge spar, with both being viewed as advantageous for rigid flapping wings. A camber forming flexibility on the wing was used, which could be varied to obtain a rigid wing configuration, in the model developed and presented here.

This concludes a review of the influence factors in achieving desired straight-and-level flight performance optimization of ornithopters. Aerodynamic mechanisms of flapping wing flight have been discussed in terms of input variables in the model. In the next section, state-of-the-art and scaling variance of bioinspired flapping wing robotics is discussed.

1.4 HISTORICAL CONSIDERATIONS FOR BIOINSPIRED FLAPPING WINGS AVIAN FLIGHT AND ROBOTICS

Smaller natural flapping wing flyers fly with a high flapping frequency at very low Reynolds numbers as compared to larger-scale natural flapping wing flyers, that fly by gliding, and with a low flapping frequency at a higher Reynolds number [20]. Larger-scale avian flyers such as albatrosses achieve range through soaring, with the wings locked in place, and interrupted by flapping motions.

Biological studies of phylogenetically similar species suggest, based on weight and lower and higher flapping frequencies, a body mass of 41 kg and a 5.1 m wingspan is the largest scale for efficient flapping wing flight. Larger-scale dinosaurs, for example, giant pterosaurs such as Pteranodon (16.6 kg, 6.95-m wingspan), Quetzalcoatlus (70 kg, 10.4-m wingspan), and Quetzalcoatlus (with a mass estimate from 85 to 250 kg) are believed to have conducted soaring flight [38]. Avian flyers encompass a transition region in the Reynolds number flight regime, where modeling efforts are challenging and results are often inaccurate. The role of the unsteady motion in the flow and its contribution to aerodynamics is not fully answered by previous studies and depends heavily on the scale of the flapping wing flyer [23]. Aerodynamic research efforts on flapping wings, which have been focused on insect flyers, are not necessarily able to be translated to avian scale flyers. Shyy states that a variation of the Reynolds number (wing sizing, flapping frequency, etc.) leads to a change in the LEV and spanwise flow structures, which impacts the aerodynamic force generation. The scaling invariance of both fluid dynamics and structural dynamics as the size changes is fundamentally difficult [6]. Table 1.1 illustrates that the difference in the Reynolds number flight regime between a fruit fly and a pigeon, with results in a comparable scaling invariance. The table provides an overview of flapping frequencies and Reynolds numbers of natural flyers.

Next, features of bioinspired avian scale vehicle platform are compared to natural flyers to further the understanding of inherent mechanisms in flapping wing aerodynamics. Aerodynamic features on birds and bats offer shape adaption to adjust the aerodynamic control surfaces [24]. Birds have feathers that are flexible, layered, and often connected to each other [24]. Bats have more than 24 independently controlled joints in the wing that enable them to fly in either positive or negative AoA [24,39]. This creates a complex three-dimensional wing topology. Birds and bats morph their wing-spans

Table 1.1 Properties of natural flyers in comparison to well-known aircraft

Species/vehicle	Mass m (kg)	Wingspan b_w (m)	Beating frequency f (Hz)	Reynolds number flight regime (Re)
Chalcid wasp (*Encarsia formosa*) [20]	2.5×10^{-8}		370	2×10^1
Fruit fly (*Drosophila virilis*) [20]	2×10^{-6}		240	2×10^2
Hummingbird (*Patagona gigas*) [20]	2×10^{-2}	0.09	15	1.5×10^4
European starling [23]	0.078	38.2	13.2	4.8×10^4
Bat (*Cynopterus brachyotis*)				
Pigeon [20]	3.5×10^{-1}		6	2×10^5
Soaring seabirds, i.e., albatross [38]	12	3	3.7–7	
Avian scale test platform	0.42–0.45	1.2–1.2	4–7	1.9×10^4 to 2.3×10^5 [9]
Stork	3.5		2	4×10^5

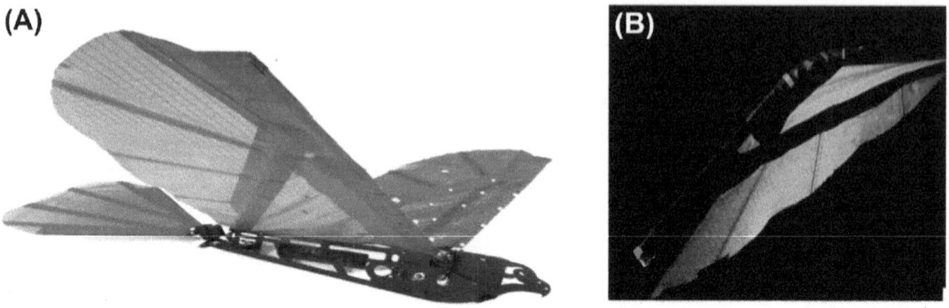

Figure 1.9 Passive wing morphing. (A) Bioinspired test platform wing morphing though thrust flap region and (B) radical shape morphing wing [15].

and change their spans to decrease and increase the wing area [24]. This speeds up the forward velocity or reduces the drag during an upstroke [24]. In both bats and birds, a slight decrease of the surface area during the upstroke occurs in fast-forward flight, while in intermediate flight speed, the surface area reduction during the upstroke becomes more pronounced. This effect is also desired in mechanical flapping wing platforms and can be achieved due to passive or active wing morphing, according to University of Maryland "Morpheus Lab" Flight Platforms (Figs. 1.9 and 1.10). These figures illustrate a passive morphing flapping wing. Passive morphing is achieved through a flexible wrist joint and a thrust flap for radical shape morphing or through a rigid leading edge spar with a thrust flap for a less radical shape change during flapping.

Birds achieve wing morphing due to feathers, which slide on top of each other, thus maintaining a smooth surface. Bat wing architecture differs and consists of a thin membrane supported by stiffening bones. This enables the

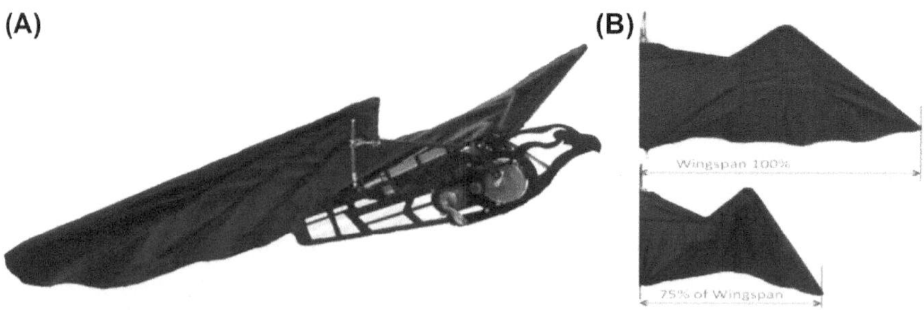

Figure 1.10 Active wing morphing bioinspired test platform. (A) Ornithopter platform. (B) Extended half wingspan versus retracted half wingspan [40].

bats to reduce their wingspan by about 20% without generating considerable slack. The membrane is elastic and can be stretched [26]. Slack in the membrane induces drag, and the trailing edge is more prone to flutter [41]. In current state-of-the-art test platforms, according to Fig. 1.10, a wingspan reduction can occur up to 25%. Fatigue and slack of the spandex membrane used in this case is an issue and could possibly be solved using smart material membranes. This would also lead to adjustable stiffness of the membrane and a tightening of the membrane during flight to avoid slack.

Bioinspired flapping wing flight platforms date back to Leonardo DaVinci's era in 1940 (Fig. 1.11). This was a human-powered ornithopter having membrane like wings. Lilienthal also developed the first successful gliding human-powered ornithopter in 1894 [22].

In the 1870s, designs of unmanned ornithopters started to appear by Jobert, Penaud, and Villeneuve [9]. The first electric powered platforms in 1961 were by Spence. More recent designs include DeLaurier's ornithopter design in 1991 from an ornithopter series started in 1975. Current ornithopter designs have a wide spectrum of performance abilities and range from the commercially available to University test platforms. Academic platforms include the "Microbat" from the California Institute of Technology, the "Phoenix" at the Massachusetts Institute of Technology, the University of Arizona Ornithopter, as well as the University of Michigan's collaborative research effort "COM-Bat," which mainly involves the collaborations of the University of California at Berkeley and the University of New Mexico [9,42,43]. Recent successful bat platforms are "Robo Bats" of North Carolina State University [1], the "Bat Bot—a self-contained robotic bat with

(A) (B)

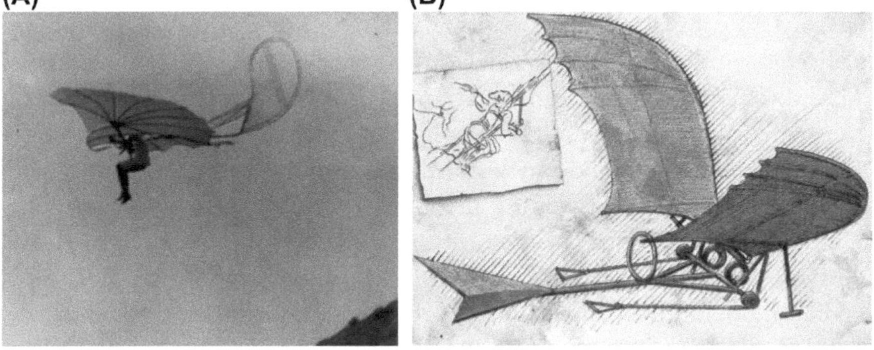

Figure 1.11 First ornithopter vehicle designs. (A) DaVincis 1490 and (B) Lilienthal 1894 [22].

soft, articulated wings" from the California Institute of Technology and the University of Illinois at Urbana–Champaign [44] and a multi-actuated Robotic bat wing of Brown University [45]. Highly successful flapping wing flight test platforms in terms of aerodynamic performance, agility, range and endurance are University of Arizona's Ornithopter Test platforms, the California Institute of Technology's platforms, and the University of Maryland Ornithopter test platforms [9,46].

1.5 OBJECTIVES IN THE DEVELOPMENT OF FLEXIBLE MULTI-BODY DYNAMICS THE MODELING METHODOLOGY DESCRIBED IN THIS BODY OF WORK

Previous sections provided the background in flapping wing flight, flapping wing aerodynamics, flapping wing robotics, and the desired capabilities of the model developed. The following will discuss the objective of modeling methodology development work and provide background on the concomitant model developed.

The objective of this work is to develop a flexible multi-body structural dynamic model considering aeroelasticity (FMBSDA) in conjunction with experimental data to identify and enhance the understanding of efficient flapping wing flight dynamics in biologically inspired ornithopters. This necessitates a novel flexible multi-body dynamics modeling methodology incorporating aeroelasticity.

The scope of this work includes creating an FMBSDA of an ornithopter by supplementing existing rigid-multi-body-dynamics models, with the inclusion of flexibility in the wing membrane kinematics; designing the validation experiments and analyzing the kinematics of an experimental ornithopter membrane wing in free flight captured through a Vicon Vision system. This provides the kinematics of the membrane wing and is used for verification of the FMBSDA model; developing FMBSDA model enhancements to obtain accurate representation of the ornithopter that is coherent with experimental data; and developing a model capable of investigation of the flight dynamics and physics of flapping wing flyers, to quantify and identify the influence of varying flexibility such as camber stiffness, wing design, and wing motion profiles, on the flight dynamics and performance and hence be used for design optimization purposes.

The modeling of flapping wing vehicles is challenging because of the complex nature of the problem as characterized by a strong coupling between fluid and structural dynamics, a nonlinear elastic multi-body system, and associated unsteady aerodynamics. Geometric nonlinearity puts the flapping wing problem into the category of nonlinear elastic multi-body systems. Flexibility in the wing membrane model is necessary for an accurate representation of ornithopter flapping wing flight dynamics. This multi-disciplinary analysis methodology requires the evaluation of tools representing individual disciplines before they can be interfaced together in a high-fidelity comprehensive model.

Disciplines like flexible-multi-body-dynamics modeling, vehicle dynamics modeling, relevant existing models of ornithopter, and flapping wing aerodynamics models are discussed in the remainder of this chapter. In addition, a comprehensive review of previously mentioned topics provides the necessary background for the model development in individual modeling disciplines involved.

REFERENCES

[1] Seeleke, Robo-Bats May Be Next Remote Control Flyers, July 7, 2009.
[2] S.S. Gheorghe Bunget, in: The 15th International Symposium on: Smart Structures and Materials & Nondestructive Evaluation and Health Monitoring, March 2008, 69282F–69282F-12.
[3] F. Bohorquez, D.J. Pines, P.D. Samuel, Small rotor design optimization using blade element momentum theory and hover tests, Journal of Aircraft 47 (1) (January 2010) 268–283.
[4] R.P. Malhan, Investigation of Aerodynamics of Flapping Wings for Micro Air Vehicle Applications (Dissertation), 2013.
[5] P. Seshadri, M. Benedict, I. Chopra, Experimental investigation of an flapping wing hovering micro air vehicle, in: American Helicopter Society Aeromechanics Specialists Conference, January 2010.
[6] W. Shyy, H. Aono, S.K. Chimakurthi, P. Trizila, C.-K. Kang, C.E.S. Cesnik, H. Liu, Recent progress in flapping wing aerodynamics and aeroelasticity, Progress in Aerospace Sciences (2010), http://dx.doi.org/10.1016/j.paerosci.2010.01.001.
[7] M. Boyer, Nano Hummingbird, February 2012 [Online]. Available: http://www.avinc.com/nano/.
[8] U. Pesavento, Z.J. Wang, Flapping wing flight can save aerodynamic power compared to steady flight, Physical Review Letters PRL 103 (2009) 118102.
[9] J. Grauer, J.E. Hubbard, Modeling and System Identification of an Ornithopter Flight Dynamics Model (Ph.D. dissertation), University of Maryland, 2011.
[10] NASA Langley. https://images.nasa.gov, 2014.
[11] Army Photo, Soldiers Maintain Security, Army Image Library, 2017.
[12] NASA Image Library, NASA ID: KSC-2012-1180. https://images.nasa.gov, January 31, 2012.
[13] Z. Jackowski, Design and Construction of an Autonomous Ornithopter (Master's thesis), 2009.
[14] M.T. Keennon, J.M. Grasmeyer, Development of the black widow and microbat MAVs and a vision for the future of MAV design, in: AIAA International Air and Space Symposium and Exposition Conference, vol. 2901, July 2003.

[15] A. Wissa, N. Guerreiro, J. Grauer, C. Altenbuchner, J.E. Hubbard, Y. Tummala, M. Frecker, R. Roberts, Flight testing of novel compliant spines for passive wing morphing on ornithopters, in: Structural Dynamics, and Materials and Co-located Conferences, April 2013.
[16] E. Avadikian, J. Hubbard, Ornithopter Development and Optimization Guide, Tech. Rep., 2010.
[17] J. Grauer, J. Hubbard, Modeling of ornithopter flight dynamics for state estimation and control, in: IEEE, American Controls Conference, June 2010.
[18] National Park Service Image Library. https://www.nps.gov/media, 2017.
[19] S.M. Swartz, M.S. Groves, H.D. Kim, W.R. Walsh, Mechanical properties of bat wing membrane skin, Journal of Zoology 239 (1996) 357–378.
[20] A. Azuma, The Biokinetics of Flying and Swimming, AIAA Education Series, Blacksburg, VA, 2005.
[21] OAI, Ohio Aeronautics Institute, Flapping Wing, Image Courtesy NASA Advanced Concepts, 2014.
[22] R.L. Harmon, J.E. Hubbard, Aerodynamic Modeling of a Flapping Membrane Wing Using Motion Tracking Experiments (Master's thesis), 2008.
[23] H. Ben-Gida, A. Kirchhefer, Z.J. Taylor, W. Bezner-Kerr, C.G. Guglielmo, G.A. Kopp, R. Gurka, Estimation of unsteady aerodynamics in the wake of a freely flying European starling (*Sturnus vulgaris*), PLoS One 8 (November 22, 2013).
[24] W. Shyy, Y. Lian, J. Tang, D. Viieru, H. Liu, Aerodynamics of Low Reynolds Number Flyers, Cambridge University Press, New York, NY, 2008.
[25] D. Floreano, J.-C. Zufferey, M.V. Srinivasan, C. Ellington, Flying Insects and Robots, Springer, Heidelberg, London New York, NY, 2009.
[26] S. Swartz, Skin and Bones: The Mechanical Properties of Bat Wing Tissue, 2010, pp. 109–126.
[27] P. Gopalakrishanan, Unsteady Aerodynamic and Aeroelastic Analysis of Flapping Flight (Dissertation), Virginia Polytechnic Institute and State University, VA, 2008.
[28] P. Ifju, M. Waszak, L. Jenkins, Stability and Control Properties of an Aeroelastic Fixed Wing Micro Aerial Vehicle, vol. 4005, AIAA Flight Mechanics, Montreal, 2001, p. A01-37226.
[29] T.J. Mueller, J.D. DeLaurier, Aerodynamics of small vehicles, Annual Review of Fluid Mechanics 35 (2003) 89–111.
[30] P. Ifju, M. Waszak, L. Jenkins, Stability and control properties of an aeroelastic fixed wing micro aerial vehicle, in: AIAA Atmospheric Flight Mechanics Conference and Exhibit, August 2001, http://dx.doi.org/10.2514/6.2001-4005.
[31] H. Garcia, M. Abdulrahim, R. Lind, Roll control for a micro air vehicle using active wing morphing, in: AIAA Paper, 2003, p. 5347.
[32] D. Lentink, M. Gerritsma, Influence of airfoil shape of performance in insect flight, in: Proceeding of 33rd AIAA Fluid Dynamics Conference and Exhibit, vol. AIAA 2003-3447, 2003.
[33] J.R. Usherwood, C.P. Ellington, The aerodynamics of revolving wings II. Propeller force coefficients from mayfly to quail, Journal of Experimental Biology 205 (2002) 1565–1576.
[34] M.A. Green, A.J. Smits, Effects of three-dimensionality on thrust production by a pitching panel, Journal of Fluid Mechanics 615 (2008) 211–220.
[35] D.L. Altshuler, R.K. Dudley, C.P. Ellington, Aerodynamic forces of revolving hummingbird wings and wing models, Journal of Zoology London 264 (2004) 327–332.
[36] S.A. Ansari, K. Knowles, R. Zbikowski, Insectlike flapping wings in the hover. Part 2: effect of wing geometry, Journal of Aircraft 45 (6) (2008) 1976–1990.

[37] C.-K. Kang, H. Aono, P. Trizila, Y. Baik, J. Rausch, L. Bernal, M. Ol, W. Shyy, Modeling of pitching and plunging airfoils at Reynolds number between 1 × 10^4 and 6 × 10^4, in: Proceeding of 39th AIAA Fluid Dynamics Conference, vol. AIAA 2009-4100, 2009.
[38] K. Sato, K.Q. Sakamoto, Y. Watanuki, A. Takahashi, N. Katsumata, C.-A. Bost, H. Weimerskirch, Scaling of soaring seabirds and implications for flight abilities of giant pterosaurs, PLoS One (April 29, 2009), http://dx.doi.org/10.1371/journal.pone.0005400.
[39] S.M. Swartz, Allometric patterning in the limb skeleton of bats: implications for the mechanics and energies of powered flight, Journal of Morphology 249 (1997) 277–294.
[40] C. Altenbuchner, Dynamic response characteristics of an active morphing ornithopter transmission, in: ASME 2011 Conference on Smart Materials, Adaptive Structures and Intelligent Systems, vol. SMASIS2011-4951, April 2011.
[41] R.M. Alexander, Mechanics of bipedal locomotions, in: P.S. Davies (Ed.), Perspectives in Experimental Biology, 2002, pp. 493–504.
[42] N.C. Moore, Birds, Bats and Insects Hold Secrets for Aerospace Engineers, March 2008 [Online]. Available: http://ns.umich.edu/new/releases/6312-birds-bats-and-insects-hold-secrets-for-aerospace-engineers.
[43] N.C. Moore, Sensors for Bat-Inspired Spy Plane under Development, March 2008 [Online]. Available: http://ns.umich.edu/new/releases/6409-sensors-for-bat-inspired-spy-plane-under-development.
[44] Robert Perkins – California Institute of Technology, Engineers Build Robot Drone That Mimics Bat Flight, February 1, 2017. http://www.caltech.edu/news/engineers-build-robot-drone-mimics-bat-flight-53794.
[45] J.W. Bahlman, S.M. Swartz, K.S. Breuer, Design and characterization of a multi-articulated robotic bat wing, Bioinspiration & Biomimetics (January 2013), http://dx.doi.org/10.1088/1748-3182/8/1/016009.
[46] D.H. Olson, D. Silin, M. Aki, C. Murrieta, J. Tyler, A. Kochevar, A. Jehle, S. Shkarayev, Wind tunnel testing and design of fixed and flapping wing micro air vehicles at the University of Arizona, in: Micro Air Vehicle Design Papers, 2005, p. 9.

CHAPTER TWO

Flexible Multi-Body Dynamics Modeling Methodology's for Flapping Wing Vehicles

2.1 CLASSIC MODELING METHODOLOGY'S

2.1.1 The Classification of Flexible Multi-Body Systems

First, a background on flexible multi-body systems is given. Originally, simple treelike topologies were handled using multi-body dynamics (MBD). The field has advanced considerably to the point that it can handle linearly and nonlinearly elastic multi-body systems as well as arbitrary topologies. A multi-body system is typically comprised of bodies, joints, force elements, and components of control.

MBD modeling is now used as a fundamental design tool in many areas of engineering [1]. MBD is used to predict forces, movements, and kinematics of multiple objects moving relative to each other and to calculate stresses and strains for elastic systems. The field of flexible MBD is concerned with the kinematic and dynamic analysis of flexible mechanical systems. Those are constrained deformable bodies that undergo large displacements to include large rotations. These large displacements are comprised of rigid-body motion as well as elastic deformations [2].

Multi-body systems can be classified as (1) rigid multi-body systems, (2) linearly elastic multi-body systems, and (3) nonlinearly elastic multi-body systems. Linearly elastic and nonlinear elastic multi-body systems fall within the category of flexible MBD models. Rigid multi-body models having flexible joints also fall within this category. Linear and nonlinear refers to the material constitutive laws and strain—displacement relationships of the flexible body. Multi-body systems are inherently nonlinear due to the large overall motions of the bodies and coriolis and centrifugal forces [1,3]. *Rigid multi-body systems* are comprised of treelike topology of rigid bodies. Rigid bodies are connected though mechanical joints and move relative to each other [1]. In this category the bodies of the systems are rigid, which means that two tracking points on the rigid body have no relative motion to each other and remain fixed in their distance. However, within the category

Modern Flexible Multi-Body Dynamics Modeling Methodology for Flapping Wing Vehicles
ISBN: 978-0-12-814136-6
http://dx.doi.org/10.1016/B978-0-12-814136-6.00002-0

lumped elastic, components may exist connecting rigid bodies, i.e., flexible joints, bushing, or force elements [1]. Flexible joints represent localized elasticity and may exhibit arbitrary constitutive behavior [1]. A rigid multi-body system schematic is presented for the ornithopter used in this work (see Fig. 2.1).

Within the category of *linear elastic multi-body models*, the systems consist of an assembly of both rigid and elastic bodies and elastic bodies only. Material constitutive laws and strain—displacement relationship of the flexible bodies remain within the linear regime. Fig. 2.2 displays an ornithopter system schematic modeled within this category. Here the flexible wings have linear elastic deformations, plain sections remain plain, strain—displacement relationships remain linear, and local deformations of the wing are small [1].

In the category of *nonlinear elastic multi-body models*, the system consists of flexible and rigid bodies or flexible bodies only. The strain—displacement

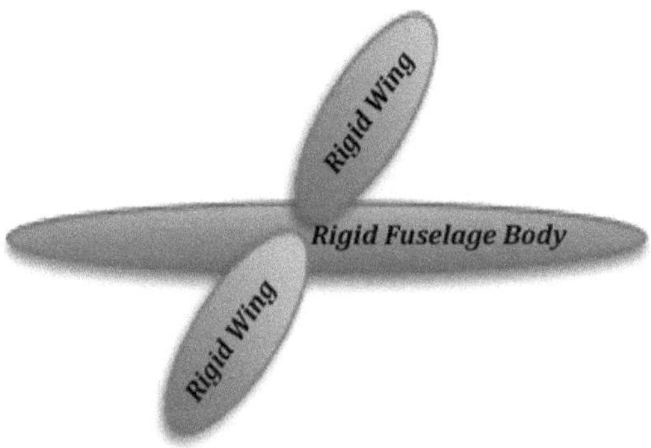

Figure 2.1 Ornithopter schematic: rigid multi-body system.

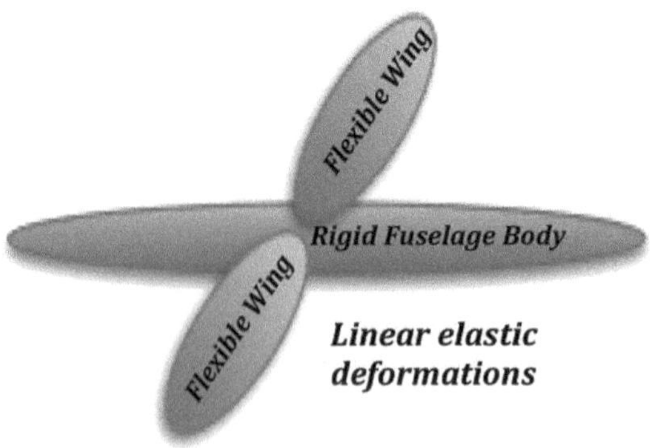

Figure 2.2 Ornithopter schematic: linear elastic multi-body systems.

relationships of the elastic bodies lie within the nonlinear regime. Modeling of nonlinear elastic multi-body systems leads to a large order of the model [4]. A mechanical system may have to be handled as a nonlinear elastic multi-body system in the case of geometrically nonlinear problems and even if linear material constitutive laws describe the material behavior adequately. For example, nonlinear elastic multi-body systems are modeled if large displacements and rotations are involved and/or if strain components become large, in case of a materially nonlinear problem, and/or a geometrically nonlinear problem (Fig. 2.3).

The nature and physics of the elastic problem need to be well understood to categorize a system with flexible bodies in the linear elastic multi-body system or nonlinear elastic multi-body system. It is advantageous to categorize a problem within the regime of linear elastic multi-body systems versus nonlinear elastic multi-body systems in terms of system order and modeling capabilities discussed in Section 2.1.2. The lines between linear and nonlinear elastic multi-body systems are sometimes blurry [1].

The categorizing of these systems presents a challenge when one considers the example of a helicopter rotor blade and wind turbine blade. As the blade rotates, elastic displacements and rotations remain small and linear material constitutive laws are adequate. In the case of a wind turbine or a slow-rotating helicopter rotor blade, linear strain—displacement relationships can be used. Using proper linear elastic multi-body systems, assumptions can predict the dynamic response accurately [1]. As the helicopter rotor blade speeds up and significant centrifugal forces occur, centrifugal stiffening and nonlinear coupling between bending and torsional deformations occur [1]. Nonlinear strain—displacement relationships must be used to capture these effects adequately. The linear material constitutive laws still

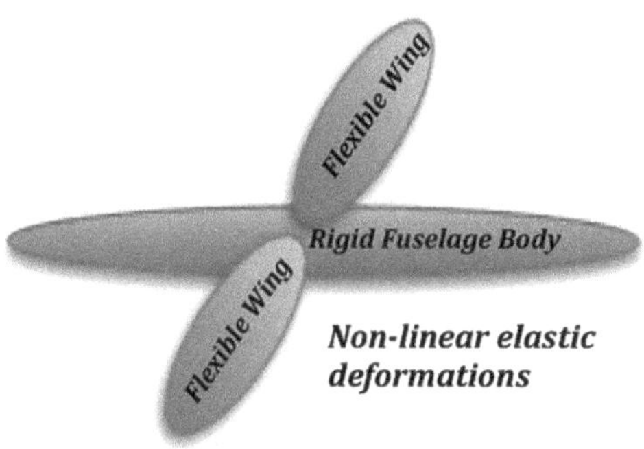

Figure 2.3 Ornithopter schematic: nonlinear elastic multi-body systems.

adequately predict the material behavior. Nonlinear elastic bodies may be present for components which operate at high-speed or high-force loads, and the nature of the problem needs to be accurately understood. Owing to aforementioned geometric nonlinearties, the problem of the helicopter rotor blade now quantifies as nonlinear elastic multi-body system [1]. Distinction between category two and three is further complicated because nonlinear elastic and linear elastic components can appear simultaneously in a system such as the ornithopter problem at hand. The dynamic characteristics of an ornithopter must be examined and understood to lead to accurate dynamic modeling. Assumptions on geometric nonlinearities and dynamic properties used in the model development here were verified and validated using its experimental in-flight kinematics and are detailed in Chapter 6.

Considering the problem of an ornithopter: as the wing flaps, elastic displacements and rotations of the wing material viewed in a wing fixed reference frame may remain very small. Therefore, linear material constitutive laws describe the elastic behavior of the wing structure adequately. The ornithopter wings are built to undergo small strains so that the structural integrity of the wing is preserved over its lifetime. Owing to the angular speed of the flapping motion of the wing, it is assumed that small centrifugal forces are induced. Considerable stiffening of the wing due to these forces as well nonlinear coupling between its two bending, torsional deformations may also remain small given the slow rotation rates. The assumption is due to the considerably low angular speed of the flapping motion and occurrent motions on the wings. Therefore, strain—displacement relationships also remain small and stay in the linear regime. Deformation assumptions are also checked in wing kinematics data and are presented in Chapter 7.

In conclusion, it is assumed that effects can be captured adequately using a linear multi-body system (LMBS) model. The dynamics of the ornithopter need to be well known to use adequate modeling techniques. All assumptions are verified to use the category of (LMBS) and are discussed in Chapter 4.

2.1.2 Flexible Multi-Body Dynamics Modeling

An overview of the categorization of flexible multi-body systems was discussed in the last section. The following section discusses and reviews the adequate and existing modeling techniques of such systems. In elastic MBD, the dynamic behavior can be calculated by a set of partial differential equations (PDEs). It is noteworthy that the solution of, for example, the bodies stress state, is analytically possible in rare cases [4]. In rigid-body systems where a body is assumed to undergo no deformation compared to the overall motion, the equations of motion (EOMs) can be derived from a set of second-order differential and algebraic equations (DAEs) [4,5]. In the case of kinematic constraints, an additional set of second-order ordinary differential equations (ODEs) or generalized coordinates must be used. Rigid-body dynamics is typically computationally fast as it has a low order of the system and a low number of degrees of freedom (DOFs). There are six DOFs per rigid body, three translational coordinates, and three rotational coordinates. The DOFs of a multi-body system can be calculated using the Gruebler Equation [6]. According to previous discussions, elastic multi-body systems are modeled by a set of PDEs, and a solution has to be approximated using the Ritz–Galerkin method, the Ritz method, the finite element method (FEM), or the boundary element method [4,5]. Depending on the nature of the problem, a combination of these methods may also be used. Under certain assumptions, the discretization procedure leads to the definition of second-order ODEs [4]. The FEM is most popular for handling elastic MBD. The FEM is most easily applicable, independent of geometric complexity, and a sparse pattern is achieved throughout the derivation of ODEs. To satisfy geometric and kinetic boundary conditions within the domain, modeling occurs using geometrical elements [finite elements (FE)] and shape functions (Ritz shape functions) for these elements [4]. The FEM method provides high accuracy for nonlinear and linear elastic structures. However, to sufficiently describe an elastic multi-body system dynamics system a large number of elastic coordinates have to be considered. This leads to a large number of FE and nodes. This is computationally expensive and in cases of large mechanical systems is unfeasible. A reduction of the DOF is desired. This can be achieved, for example, by the handling of underlying rigid-body motions and a superimposed elastic deformation. Significant inertia changes due to elastic deformations may also occur, which leads to a coupling of the underlying rigid-body EOMs and the structural dynamics EOMs. Therefore, these EOM's cannot be solved independently. Nonlinear structural bodies within the system as well as aeroelasticity also lead to coupled EOMs. Coupled and uncoupled methods used to integrate rigid and elastic dynamics exist for such hybrid systems. The basic approaches used in simulation and computer-aided calculations for flexible MBD and kinematics are the floating frame of reference formulation, the finite element (FE) incremental method, large rotation vector formulations, the finite segment method, and the linear theory of elastodynamics. In general, modal expansion methods are reliable within the regime of linear elastic multi-body system [1]. MBD analysis is not yet widely applicable when dealing with nonlinear elastic systems, and a deep understanding of the system model is required [7]. A more detailed review development in the field of flexible MBD and related methods can be found in Refs. [1,2,4] and is further discussed in Chapter 4.

2.1.3 The Implementation in Available Code and Software

A brief overview of the theoretical background of modeling of flexible multi-body systems was presented in previous section. The state of the art, the choice, and integration of solvers for the EOMs are discussed in the following section. Several available codes and software packages exist to handle flexible MBD problems. These are commercially available and through university research codes. Codes and software packages incorporated in this work were chosen based on the ability to incorporate multi-physics, change the solution procedure by adjusting code, as well as to couple aerodynamics though both code and computer-aided computational fluid dynamics (CFD). Due to the complexity of the ornithopter problem and the experimental verification methodology described in Section 2.2, the solution procedure is required to be versatile and customizable through code. The following are the model capabilities and requirements: (1) ability to appropriately model geometry though computer-aided design (CAD), functional building blocks or coding, (2) handle rigid and flexible MBD linear and nonlinear adjustable solution procedures, (3) ability to use interface for aeroelasticity.

Flexible MBD modeling has generated large interest due to its versatility and applicability and is a growing field. The modeling capabilities are desired in many areas of engineering such as energy, automotive, and aerospace. For example, wind turbines, helicopter rotor blades, space manipulators, and solar sails lead to a wide array of existing codes and model assumptions and simplifications. Fidelity problems for nonlinear elastic MBD applications lead to constant advancements and the evolving of codes and software packages. In recent years, fast advancing and quick turnarounds in software packages have been made to handle and categorize elastic MBD problems. A comprehensive list of evolving software and codes available can be found in Table 2.1.

For aerospace applications, the academic code, CAMRAD, was developed for helicopter and rotorcraft dynamics and aerodynamics and is commercially marketed by analytical methods. It includes MBD, nonlinear FEs, structural dynamics, and rotorcraft aerodynamics. CAMRAD provides excellent performance abilities for rotorcraft applications but is not well suited for the ornithopter problem due to its limited versatility [8]. MBD

Table 2.1 Finite element method (FEM) and multi-body dynamics (MBD) university codes and commercial packages

Code/software	Primary type	Geometry	Rigid MBD	Flexible MBD	FE linear	FE linear and nonlinear	AEOI	Developer
ADAMS solver	MBD	Yes CAD	Yes	Yes linear	Yes	No	No	MacNeal-Schwendler Corporation
MARC solver	FEM	Yes CAD	No	No	Yes	Yes	No	MacNeal-Schwendler Corporation
DYMORE	MBD	Programmable toolbox	Yes	Yes	Yes	Yes	Yes	Georgia Tech University
PATRAN/ NASTRAN	FEM	CAD	No	No	Yes	Yes	No	Originally NASA marketed by MSC
ABAQUS	FEM	CAD	No	No	Yes	Yes	Limited	Dassault Systems
ANSYS	FEM	CAD	No	No	Yes	Yes	Limited	ANSYS
CAMRAD	MBD	Helicopter geometry	Yes	Yes	Yes	Yes	Yes	Johnson Aeronautics
COMSOL	MP	CAD	Yes	Yes	Yes	Yes	Limited	COMSOL
SIMPACK	MBD	CAD	Yes	Yes	Yes	Yes	Limited	Originally DLR marked by SIMPACK AG and MathWorks
LS-DYNA	MBD	Programmable toolbox	Yes	Yes	Yes	Yes	Yes	Livermore Software Technology Corporation
MBDyn	MBD	Programmable toolbox	Yes	Yes	Yes	Yes	Limited	Polytechnic di Milano

AEOI, aero elastic open interface.

codes for aerospace applications, such as MBDyn and developed by the Polytechnic di Milano and DYMORE by Georgia Tech University, were also considered for this work. MBDyn features the integrated multi-disciplinary simulation of multi-body, multi-physics systems, including nonlinear mechanics of rigid and flexible bodies and can be coupled with aerodynamic codes and CFD. DYMORE is a finite element based tool for the analysis of nonlinear flexible multi-body systems [7]. COMSOL is a multipurpose software platform, which provides a multi-disciplinary solution capability but is ill-suited for the MBD simulation [9]. Individual discipline software packages were considered here for integration into a multi-physics simulation through an interface for high modeling fidelity. PATRAN/NASTRAN, ANSYS, and ABAQUS were considered to handle the structural dynamics. PATRAN is a pre/postprocessing software for finite element analysis (FEA) by the MacNeal-Schwendler Corporation (MSC). NASTRAN is a multi-disciplinary structural analysis (FEA) application for PATRAN. PATRAN/NASTRAN marketed by MSC was originally developed for NASA in the late 1960s. ANSYS and ABAQUS are other established and the most widely used commercially available FEA software tools. ABAQUS is known for its high-fidelity nonlinear structural modeling capabilities. MARC is a nonlinear FE solver by MSC and also suitable to handle MBD of the problem. LS-DYNA is a general-purpose FE program capable of simulating nonlinear dynamics and rigid-body dynamics, mostly used for rotor blade applications. SIMPACK is another possible simulation software for general-purpose multi-body simulation. It is especially well suited to high-frequency transient analyses. SIMPACK is a commercially available package and was primarily developed to handle complex nonlinear models with flexible bodies and harsh shock contact originally by the German Aerospace Center (DLR). There are limitations in aerodynamic interfaces with SIMPACK. MSC Advanced Dynamic Analysis of Mechanical Systems (ADAMS) is the world's most widely used MBD software, and ADAMS solver along with the MARC solver used for modeling the MBD component of the problem. Solvers are facilitated in the multi-disciplinary SimXpert Graphical User Interface (GUI). Interface of ADAMS solver with previously mentioned FEA tools is possible through a mode neutral files. PYTHON, MATLAB, and FLUENT were considered to handle the latter analytical aerodynamics or CFD for the fluid–structure interaction (FSI). The choice of programming language or software for the aerodynamics is based on the related flapping wing aerodynamic models discussed in Section 2.1.6 where proven theories were utilized and extended in this work. PYTHON and MATLAB are widely used programming languages. ANSYS FLUENT is a commercially available CFD software package in which preceding flapping wing aerodynamic modeling research was carried out. Owing to the incorporation of experimental kinematics, aerodynamics was modeled and computed using MATLAB and also interfaced through SimXpert.

SimXpert released in 2010 by MSC was chosen as the GUI for the multidisciplinary simulation methodology. Within the frame of this effort, it allows interfacing with MATLAB for an aeroelastic consideration of the problem and allows the ability to be interfaced with CFD for a higher fidelity, yet more costly simulation in the future. A versatile use of ADAMS solver for MBD or an explicit nonlinear solver utilizing LS-DYNA opens up the work space for the flexible multi-body simulation. Linear and nonlinear structural dynamics of bodies in the multi-body system can be calculated using MD NASTRAN solver integrated in the GUI. The geometry of the ornithopter test platform is modeled using PTC Creo CAD software and is imported into the SimXpert GUI. For all simulation computations, a multi-disciplinary software SimXpert GUI was employed. The solution procedure used here is according to Fig. 2.4, and a fully integrated model image is shown in Fig. 2.5.

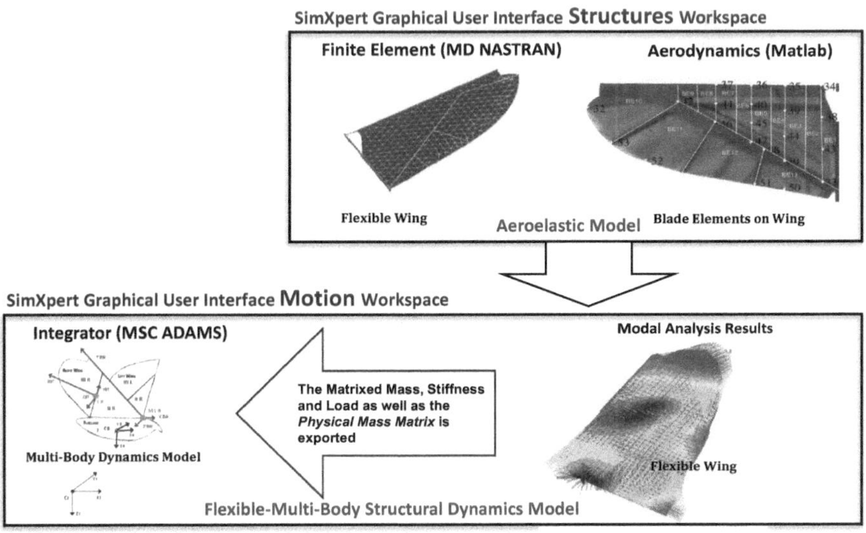

Figure 2.4 SimXpert: flexible multi-body dynamics model implementation.

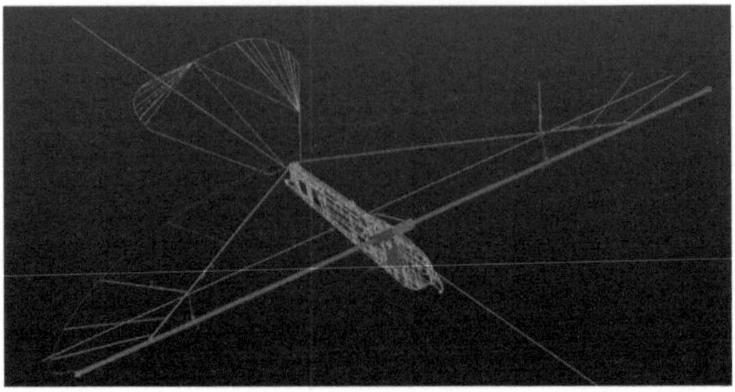

Figure 2.5 SimXpert: Image of fully integrated flexible multi-body dynamics model of ornithopter.

Aeroelasticity was incorporated by a FE computation of time-variant (TV) aerodynamic loads as applied to the structure. Time-variant loads (TVL) result from kinematics data described in the following chapters. For the determination of the time histories and magnitudes of TVL, MATLAB was used. FE and MBD calculations were performed using MD NASTRAN solver and MD-ADAMS motion solver, respectively [10]. The aeroelastic analysis provides mass, stiffness, and load matrixes, as well as the physical mass matrix and is exported into the MBD integration. A semiphysical modeling methodology is used which is further explored in Chapter 4.

2.1.4 The Vehicle Dynamics Modeling of Ornithopter

The level of complexity of vehicle dynamics models varies from simple single rigid-body models to more complex rigid MBD models, and finally the most complex type of flexible MBD models. These elastic effects can also couple with the aerodynamics associated with the system, which makes it an aeroelastic problem. This section discusses vehicle dynamics modeling in the context of ornithopter modeling and provides a review of shortcomings of related models. Fig. 2.6 shows the complexity of models on an example of an ornithopter.

A variety of models that describe the flight dynamics of flapping wing flyers have been previously developed. Several rigid MBD models for flapping wing vehicles have been derived using an array of methods [3,11–14]. A single rigid-body dynamics model according to standard vehicle dynamics is shown in Fig. 2.6A. These standard aircraft EOMs can be mathematically described by the evolving position and velocity states of the single rigid body. Here the inertial effects of wings are negligible,

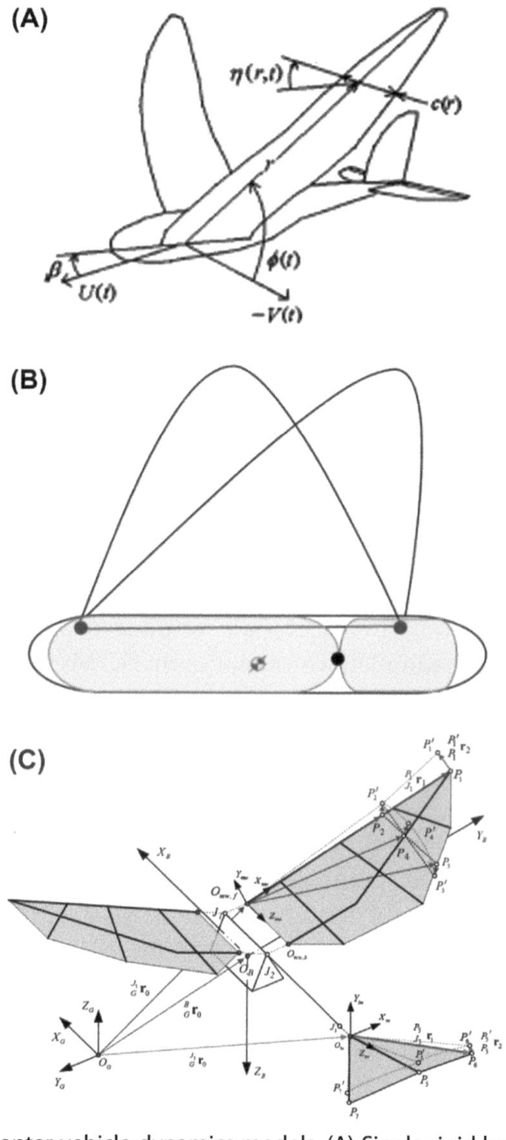

Figure 2.6 Ornithopter vehicle dynamics models. (A) Single rigid body, (B) rigid multi-body dynamics, and (C) flexible multi-body dynamics model [11].

and flapping frequencies are significantly separated from dynamic frequencies [3,15].

A single body representation is a commonly used modeling methodology for small-scale ornithopter problems but in limited cases valid for avian scale ornithopter [3,16,17]. Conventionally, single body dynamics models for aircraft are nonlinear models in a body dynamics sense [3,15]. Mass distribution variations due to flapping wings are significant here for the ornithopter test platform and the moments of inertia of the wing can vary up to 53.6% [3]. Dynamic frequencies are also in close proximity to flapping frequency and require a MBD representation of vehicle dynamics models for ornithopter. These models are more complex and they contain position and velocity states for the individual rigid bodies out of necessity, as they can capture the inertial effects of the flapping wings [3]. An investigation by Orlowski employed a multi-body model and empirical scaling laws observed in nature. Their research concluded that linear momentum effects from flapping wings are always significant. Angular momentum effects are also significant for flapping frequencies below 40 Hz, and magnitude decreases below with rising flapping frequency up to 40 Hz [3,18].

Fig. 2.6B shows a multi-body model of an ornithopter where wing bodies and fuselage bodies are treated as rigid bodies. A version of the rigid MBD model of an insect has been previously developed by Dyhr [12], and a sketch of it is highlighted in Fig. 2.6B. Fuselage (abdominal) body dynamics are of interest and aerodynamic forces are introduced, which are linear time invariant (LTI) into the models' constraint term. EOMs for a model inspired by the Hawkmoth are linearized about a hovering equilibrium.

Several other models have been employed using time averaging theory to simplify the periodic forcing of the wings in a similar manner [18–20]. A common objective for the utilization of flapping wing modeling is for stability analysis and control of these vehicles. A number of models have been developed for these purposes. To most consistently employ classical and modern control, linear perturbation models are often desired for the stability analysis and controls synthesis. LTI models result in equilibrium points with Eigenvalues and Eigenvectors [3]. Faruque and Humbert also developed (LTI) models of insects using system identification techniques. They developed a reduced-order model of longitudinal hovering flight dynamics for dipteran insects. Here a quasi-steady wing aerodynamics model is extended by including perturbation states from equilibrium and paired with rigid-body EOMs to create a nonlinear simulation of a *Drosophila*-like insect [13,21]. The fidelity of LTI models for most control purposes and design optimization purposes of interest in this work are questionable even by the authors who have developed them [14,22]. As a consequence Grauer, J. Lee, and Taylor have developed more complex models [11,14,23]. Taylor developed a semiempirical model of the longitudinal fight dynamics of desert locusts *Schistocerca gregaria* by representing approximate periodic instantaneous forces using Fourier series embedded in the EOMs resulting in a nonlinear time-periodic model [14]. Grauer developed a rigid MBD model that uses energy methods to determine EOMs and cast them into a canonical form [3]. The model uses a TV aerodynamics model identified through system identification techniques. It was primarily developed for control purposes and recommended including wing flexibility to result in a higher fidelity for dynamic investigations. Fig. 2.6B (highlighted portion) shows the schematic of a three-body rigid multi-body model developed by Grauer through system identification techniques to capture dynamics of experimental research ornithopter [3]. According to the previously discussed vehicle dynamics models of category, Fig. 2.6A and B mainly serve stability and control purposes.

Flexibility in the wings is a significant design metric to improve the performance of flapping wing vehicles. Flexibility also influences the stability and control characteristics of aircraft [24,25]. Wing flexibility significantly increases the level of complexity of an ornithopter model, its DOF and entails an aeroelastic response. A kinematics model for a platform that features bat-inspired wings with a number of flexible joints was completed by Bunget et al. [26]. Sparse literature exists regarding the flight dynamics of ornithopters, flexibility and FSI, and MBD [11]. Kim et al. state that generally, for the analysis of flight dynamics of ornithopters, complex nonlinear flexible multi-body configuration of an ornithopter is simplified to linear rigid-body dynamics. Some relevant ornithopter dynamic models developed by Dietl and Garcia use a single rigid-body model [27,28]. Further relevant rigid-body dynamics models are developed by Rashid and Bolder. Rashid developed a rigid-body flight dynamics model of a full-scale ornithopter, and Bolder et al. built a rigid MBD model for flapping wing Micro Areal Vehicles (MAVs) using Kane's equations [29]. In particular, the passive deformation of a flexible wing structure is oftentimes not considered or at best assumed to have a prescribed form to guarantee enough lift and thrust to propel the vehicle aloft [11,12,23,27]. Limited related research considering aeroelastic vehicle dynamics of flapping wing vehicles are further discussed in the next section followed by a review of related aerodynamic models of avian scale flapping wing.

2.1.5 An Aeroelastic Analysis of Flapping Wing Vehicles

J. Lee et al. developed a version of a flexible MBD model of an ornithopter considering aeroelasticity and represents the only preceding model developed in open literature addressing MBD, flexibility, and aeroelasticity for avian scale ornithopter (Fig. 2.6C) [11,30]. The wings are treated as flexible bodies and the wing structure is modeled using a finite element method and a superimposed flapping motion. This model has shortcomings in verification and is based on the assumption that an entire wing body stays within a linear regime. Analysis of free flight test data presented in this work confirms this is not the case for state-of-the-art test platforms. A multi-body representation of an individual wing has to be used to accurately use these assumptions. Verification of geometric nonlinearity's wing deformation and coupling between torsion and bending deformation is necessary to verify modeling assumptions. The flexible multi-body model by J. Lee provides limited experimental validation and uses a loosely coupled aerodynamic model [11]. Due to the complex nature of the problem to successfully develop the proper analysis tools, it is imperative that the model be validated and developed in conjunction with experimental flight data exposing wing motion profiles and actual dynamic performance. Furthermore, coupling between elastic coordinates and underlying rigid multi-body motion is solved independently and not in a coupled fashion. This limits model fidelity and utilization for design parameter optimization.

A comprehensive paper addressing modeling of aeroelastic vehicles, in general, can be found by Waszak and Bruttrill [25]. All aeroelastic vehicles can be described within two categories. In the first category, the coupling term between the elastic body and the aerodynamics is due to the aerodynamics only. In the second category, the inertial coupling also occurs due to elastic deformation of a flexible body [25]. Inertial effects due to the elastic displacements are uncoupled in Lee's model [11,30]. The coupling of elastic and rigid-body modes is advantageous to provide ornithopter configuration versatility by keeping modeling fidelity [11]. The model developed here considers inertial coupling due to elastic displacements beside the underlying rigid body displacements. Elastic displacements are also considered for the calculation of aerodynamic forces and results in an enhanced aeroelastic calculation.

The shortcomings as discussed in relationship to the vehicle dynamics model of flapping wing can be summarized as having one or more of the following issues:

1. Flexibility in the wings are neglected [3,23,27,29,31].
2. Vehicle dynamics model valid for one configuration only [3,23].
3. FSI is neglected or not included in model fidelity [3,27,29,31].
4. Only existing vehicle dynamics model addressing wing flexibility, multi-body and FSI by J. Lee [11] is limited though.

a. Limited experimental verification.
b. Simplified wing kinematics and flexibility: Flexibility assumptions are not in tune with flexibility and wing deformation kinematics found in working ornithopter platforms.
c. Inertial coupling through elasticity is neglected.
d. Limitations in accuracy of FSI (1) coupling methodology: loosely coupled aerodynamics model, (2) aerodynamic model assumptions itself, (3) input in aerodynamic model: resulting aerodynamic loads are based on simplified deformation assumptions of a flapping wing in free flight and not on experimental wing kinematics.

These shortcomings are addressed in the model developed in this work. This completes the review of ornithopter vehicle dynamics models and related models. Relevant flapping wing dynamics for the aerodynamic component in the multi-physics problem is reviewed later.

2.1.6 Related Avian Scale Aerodynamics and Models

A variety of analytical, experimental, and numerical models have been developed to calculate aerodynamic forces on the flapping wing. In addition to forward motion that fixed wing experiences, a flapping wing also flaps up and down and can deform a wing membrane to adjust its camber. For example, a bird can sweep forward and back, twist along the span, and fold to adjust the wing platform area [32–34]. Owing to the resulting unsteady low Reynolds number, flow and circulatory effects, or a combination of these capabilities, the accurate prediction of the flow field is challenging. The Reynolds number of the modeled test ornithopter presented here varies between 19,000 and 232,000 for steady-level flight, which encompasses the transition region where flapping wing vehicles are more aerodynamically efficient than fixed wing vehicles [3].

The most popular analytical model for avian flyers developed by DeLaurier is a quasi-steady blade element model to capture wake/vortex interactions, poststall phenomenon, and partial leading edge suction [35]. DeLaurier used modified strip theory (MST) approach shown in Fig. 2.7 to develop a quasi-steady blade element model.

DeLaurier's model is based on the modified Theodorsen function, or three-dimensional Theodorsen function that takes into account a three-dimensional wake correction factor in terms of the wing aspect ratio and the reduced frequency [11,35]. Contributions of the wing section (shown in Fig. 2.7) mean angle of attack, camber, partial leading edge suction, and friction drag are taken into account here. This allows the model to be

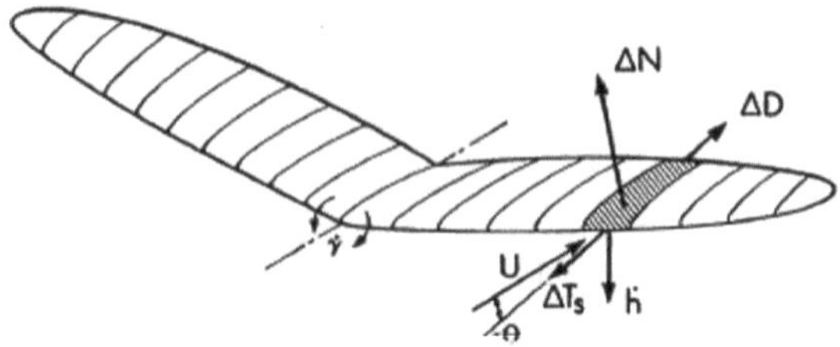

Figure 2.7 Quasi-steady blade element model [35].

used for the calculation of the average lift, thrust, power required, and propulsive efficiency of a flapping wing in equilibrium flight [28]. The wing's aspect ratio is assumed large enough that the flow over each section is essentially chordwise and therefore in the mean-stream direction according to observed avian scale flyers [36]. Variants of this model are widely used in the literature of Jun-Seong Lee who tested the design-oriented aerodynamic model for unsteady aerodynamics proposed by DeLaurier using flight tests of avian scale ornithopter [28]. He concluded that the model failed to predict the amplitude of the time-varying aerodynamic forces and moment according to the continuous flapping wing motion. The model is limited so as to classify the attached flow range and the poststall flow range [11,35].

A modified version of DeLauriers blade element theory (BET), namely MST stall has been developed by Kim et al. The improved version of the MST includes a dynamic stall model for large plunging wing motion. Here the dynamic stall cut off angle is adjusted to a flapping and plunging plate. DeLaurier's analytical models dynamic stall cut off assumption rely on data for a fixed wing. Here the inclusion of dynamic stall in the model shows significant improvement in correlation with the root mean square values of aerodynamic coefficients to experimental data of a flapping plunging plate by Okamoto [30,37]. Jae-Hung's assumptions showed improvements compared with DeLauriers original model and are adopted in current research. The classification of the flow is more refined in terms of attached flow range, dynamic stall range, and poststall flow range.

However, the model fails to take into consideration the 3D wing topologies of flexible wings, namely flexibility adjustments in the chordwise direction taking fore-and-aft camber into consideration. More refined flight kinematics obtained from experiments shows the necessity of this adjustment [38].

Walker studied the unsteady aerodynamics of deformable airfoils and used a series of Chebychev polynomials to represent the deformation of the airfoil. The aerodynamics model was developed using potential flow aerodynamics and the Joukowski conformal mapping technique was used to map the flow. He concluded that Theodorsen's function reduces the magnitude of the lift due to circulation as reduced frequency increases and shifts the phase of the circulatory lift with respect to the motion for all airfoil deformation. The magnitude of the phase shift is dependent on the reduced frequency [39]. An analytical model using BET has been previously developed in the framework of a research effort of the University of Maryland Morpheus Lab [38]. The analytical model has been developed in conjunction with experimental data in parallel with a CFD model [40]. An analytical BET developed by Harmon uses refined wing kinematics of experimental ornithopter platform and was defined by quasi-steady motions, including flapping or plunging, pitching, and forward motion [38]. However, the effects of unsteady flow mechanisms, such as leading edge suction, wake capture, dynamic stall, or spanwise flow were neglected. The momentum theory with an actuator disk that is the swept area of the wing was used to adequately capture induced velocity magnitudes on the wing. The core value of this model is defined by its development with experimental wing kinematics, which serves as an input variable and allows a TV aerodynamic force generation on the blade elements and is therefore spatially distributed over the wing. The model was developed for the membrane behavior of a flexible experimental wing ML101 and MSK004 test platform (Chapter 3) and was measured in clamped conditions. Here the membrane behavior is computed and divided in a fore-and-aft approach for two chordwise regions of the wing. The upstream region is modeled using first-order modeling assumptions that have a pitching axis located at the chord point with a time-varying length of the blade element. The blade element length of the chordwise downstream region is fixed and takes no flexibility into account. Attached flow is assumed regardless of the relative angle of attack. Kinematic fluctuations of the camber during stroke transition are neglected, which is positive during downstroke and negative during upstroke. In the downstream chordwise region the inflow angle for blade elements is adjusted by the pitch angle of the blade elements that are in the upstream region. When compared with CFD, the blade element model was determined to provide a good approximation of the complex problem of flapping wing aerodynamics and at a reduced computational cost [38,40–42]. This work concludes that the quasi-steady circulation model by Harmon can capture the aerodynamic behavior when adjusted to account for unsteady motion and membrane aerodynamics [38]. With a rising reduced frequency the unsteady effects become more pronounced as the influence of the wake is increased. The wake couples aerodynamic forces from one time step to another. Therefore the wake is a significant factor in aerodynamic force generation and increasingly depends on the wing shape of the previous time step as the reduced frequency increases. Wake structure has to be included in modeling to archive the desired fidelity. Vortex lattice methods (VLMs) provide the means to do so and is a simple numerical method to compute lift and drag. The VLM neglects the viscosity of the flow field and has been used to analyze flapping wings by several authors but provides moderate fidelity [36,43].

For example, Stanford et al. developed an unsteady model by the means of the VLM. The model results in the derivative of time-averaged vertical force, propulsive force, power, and propulsive efficiency of the flapping wing at the time step. The model was employed for a gradient-based design optimization but showed limited fidelity [36,43].

To obtain viscous solution, CFD simulations have been carried out by numerous authors. Mainly flapping and plunging structures without consideration of flexibility is addressed in the literature. Consideration of 3D effects is also limited. Unsteady aerodynamics of Micro Air Vehicles were investigated using a Reynolds Averaged Navier—Stokes (RANS) solver to compute a pitching and plunging and compared with other experiments and results correlated well (RTO, 2010) [44]. These studies agree that 3D effects have to be considered for flapping wing platforms with low aspect ratios for design optimization purposes. A study by Visbal concluded comparing a 3D and a 2D computation of the instantaneous spanwise vorticity component at a given phase of a pure plunging airfoil that 2D and 3D effects are in close agreement over a significant portion of the airfoil. A study was also carried out for an SD7003 airfoil section using large-eddy simulation (4 degrees angle of attack, Reynolds number in the order of 10^4, and a reduced frequency of $k = 3.93$) [45].

One can note that the significance of 3D effects depends on the aspect ratio, Reynolds number regime, and reduced frequency while the leading edge vortex formation exhibits a well-defined two-dimensional character. The study by Visbal concludes that as the Reynolds number increases above 4×10^4, differences between -D and 3D results become significant because coherent vortices observed in the -D simulations break down in spanwise direction [36,45]. The Reynolds number of interest in this modeling effort varies between 1.9×10^4 and 23×10^4 and a consideration of 3D effects becomes beneficial. Below a Re of 4×10^4 the flow is essentially laminar given the case study of Visbal et al. Minor differences are observed near the trailing edge aft portion and in the near-wake due to translational effects, and the 2D approach predicts the flow well in this regime. In this dissertation developed cases represent a transition region in the Reynolds number flow regime where 3D effects become significant. The previously mentioned study by Yuan et al. overpredicts force generation using a three-dimensional approach and overpredicts force generation when compared to a two-dimensional approach [46]. Here a highly three-dimensional flow was observed for the root flapping case.

In summary, authors note that 3D effects of low aspect ratio wings have to be considered for aerodynamic force prediction [46]. Most studies considering viscous flow have been carried out using 2D CFD analysis on pitching and plunging rigid airfoils by the means of nonlinear flow equations or the Navier–Stokes Equations [47,48]. A variety of further CFD codes have been developed for flapping wing [40,45,46,49]. Shortcomings include limited experimental verification and validity for avian scale ornithopter. The inclusion of flexibility in CFD studies is rare as a CFD prediction of avian flight with flexible wings presents many difficult challenges such as large elastic grid deformations and the possibility of laminar to turbulent transition [40]. Generally CFD is computationally expensive for FSI. A summary of the most applicable and successful studies for the problem at hand is discussed in the following.

Roget et al. computed aerodynamic forces produced by a flexible flapping wing and were simulated using a CFD-based methodology and compared with integrated forces obtained from experiment. The CFD model is based on the compressible Reynolds averaged form of the Navier–Stokes equations and uses a structured curvilinear grid. Refined experimental wing kinematics from experiment were used as input variable and an efficient grid deformation algorithm is devised, which deforms the body-conforming volume grid at each time step consistent with the wing boundaries (Fig. 2.8).

The study was previously conducted using MSK004 and ML001 ornithopter test platform at a Reynolds number regime of 10.5×10^4–21.3×10^4 using a modified version of University of Maryland unsteady Navier–Stokes solver (UMTURNS). The deforming grid methodology was adjusted to the flexible wing with large deformations. The UMTURNS codes are a compressible RANS solver, which uses body-fitted curvilinear structured grids. Grid deforming methodology was adjusted to accommodate the flexible wing with large deformations. In the model, the 2D planes along the wingspan (Fig. 2.8) extend

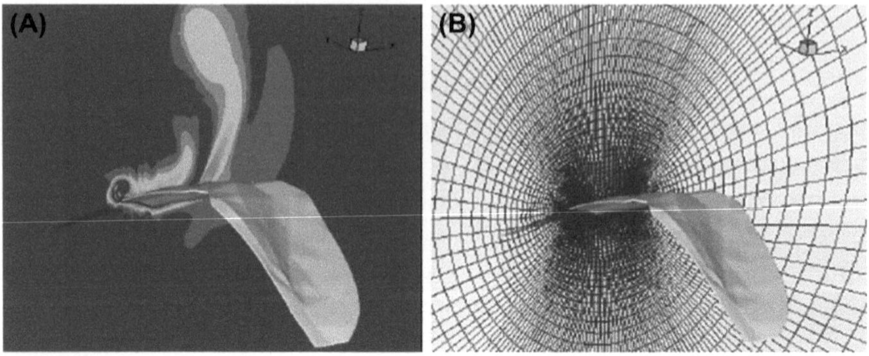

Figure 2.8 CFD analysis showing (A) vorticity contours on the flexible deformed wing and (B) deformed grid at 50% span location during upstroke of the wing at start of upstroke [40].

in the normal direction and a cosine interpolation is applied to form a three-dimensional grid that is valid in deformed and unreformed positions of the flexible wing [40]. Fig. 2.8 shows a plane of two-dimensional deformed grid and vorticity contour results of the computation on a deformed flexible wing.

The data correlation of simulation results and experiment showed good agreement with measured vertical force and satisfactory agreement with measured horizontal force at low flapping frequency. Prior work concludes that a higher fidelity of model prediction with decreasing flapping frequency and suggests inertial effects from the vehicle body and support mounts may be one of the contributors to disagreement between data and analysis [40]. FSI coupling of the solver developed by Roget et al. to the flexible MBD solver is computationally expensive. Grauer et al. developed an aerodynamic model of an avian scale ornithopter MSK004 test platform using system identification and flight test [3]. The model adequately predicts aerodynamic force generation of ornithopter wings by means of aerodynamic coefficients. The model structure determination was performed in the time domain with equation error and stepwise regression [23]. The model structure consists of aerodynamic coefficients in nonlinear expansion in terms of the state variables. Fig. 2.9 shows the rigid body kinematics fit in flight test data used for the model structure determination through system identification.

The validity of the model is limited to flight test configuration and not suited for FSI. It adequately takes into consideration the flight test occurrent flow physics due to the development through flight test data and can be used

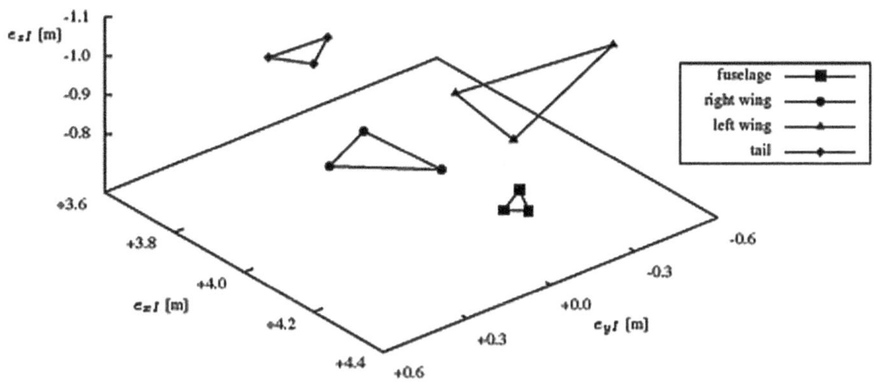

Figure 2.9 Rigid-body ornithopter fit in flight test data for aerodynamic model structure determination [23].

as the configuration specific verification point for the aerodynamic model development addressed in this work.

A previously mentioned efficient aerodynamic model suggested by Kim et al. was based on DeLauriers MST and further improved to take into account a high relative angle of attack for dynamic stall effects induced by the pitching and plunging motions. Limited experimental validation was provided with experimental data of a rigid rectangular wing but showed improvements compared with the MST and were therefore adopted in this work [50].

Although these studies were very insightful, further considerations are required to gain a complete understanding of the complex aerodynamic phenomena during actual flight. An approach of combining previous insights and modeling approaches, integrating them in a high-fidelity and low computational cost model for FSI is taken into account, in addition to considering experimental in flight kinematics.

In summary, previously developed analytical, experimental, or numerical (CFD) aerodynamic analyses mentioned earlier suffered from one or more of the following shortcomings or limitations:

1. Simulation and modeling of flapping plunging plates without the inclusion of flexibility [45–49].
2. Most studies were carried out in a very low Reynolds number flow field regime, which can be contributed to MAV scale and not avian scale ornithopter, which is of focus here [16]. This has implications based on considerations mentioned in previous sections.
3. Limited experimental validation,

a. If experimentally validated mainly through Particle Image Velocimetry (PIV) of rigid flapping and plunging plates without the consideration of flexibility, which considerably alters the flow structure, Reynolds number flight regime in the scale of MAV's [16,39,46,49].
b. Limited experimental in flight verification, one study exists by J. Lee at al. but does not consider wing kinematics and is merely justified by flight trajectory force integration [11].
c. If experimental wing kinematics are considered wake effects, which are significant for the unsteady flow field and aerodynamic force production are neglected and provide limited fidelity [38].
d. Verification is limited to constraint data measurements through PIV or Vicon, which significantly influences circulatory effects and pressure distribution on the wing (the aerodynamic force generation) [40].

4. Computationally expensive CFD simulation without the consideration of FSI or inertia effects that actually occur in avian scale ornithopter [40].
5. Aerodynamic model though system ID is verified though flight test but tailored to one specific ornithopter configuration and therefore not suitable for design optimization purposes [23].
6. 3D effects or spanwise lift distribution are neglected [38,51].

The effort presented here combines and improves on successful models previously developed to negate their individual shortcomings. Refined experimental wing kinematics are used for model development, the inclusion of wing flexibility, wake effects, dynamic stall, and spanwise lift distribution is considered in a computationally efficient manner by using BET. The model is tailored to the avian scale Reynolds number flight regime. The model is experimentally verified through flight test and inertia effects, and FSI is considered through coupling with flexible MBD model.

2.2 MODERN MODELING METHODOLOGY

The workflow overview of the model development is presented in the following chapter. This results in the novel flexible MBD modeling methodology. In Figs. 2.10–2.16, the experimental input is marked in blue. Red represents the model and yellow represents the experimental validation.

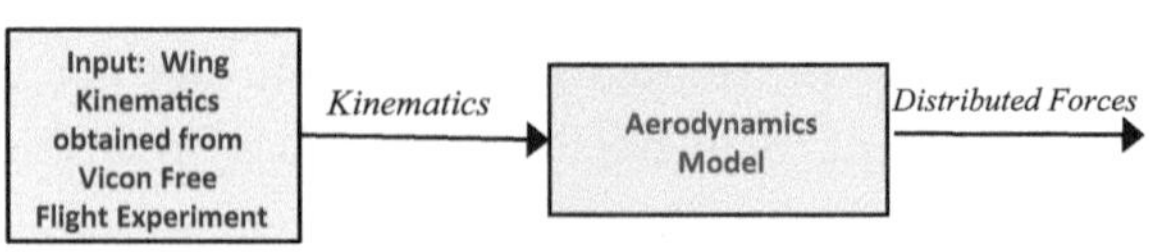

Figure 2.10 Modeling methodology: workflow stage 1.

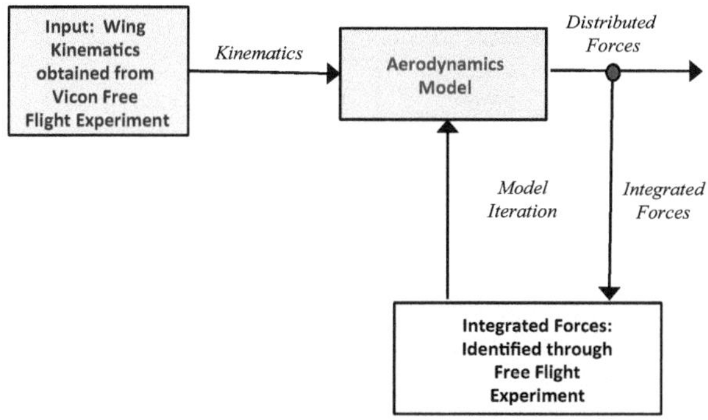

Figure 2.11 Modeling methodology: workflow step 2.

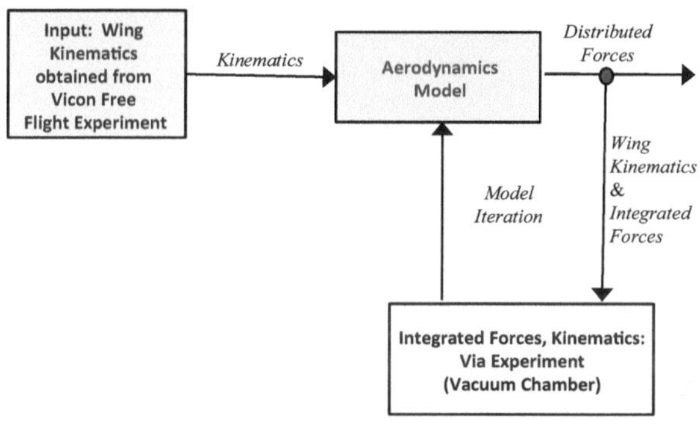

Figure 2.12 Modeling methodology: workflow step 3.

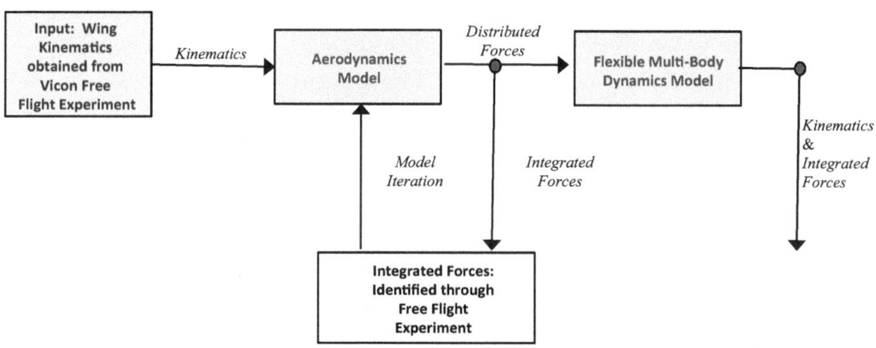

Figure 2.13 Modeling methodology: workflow stage 4.

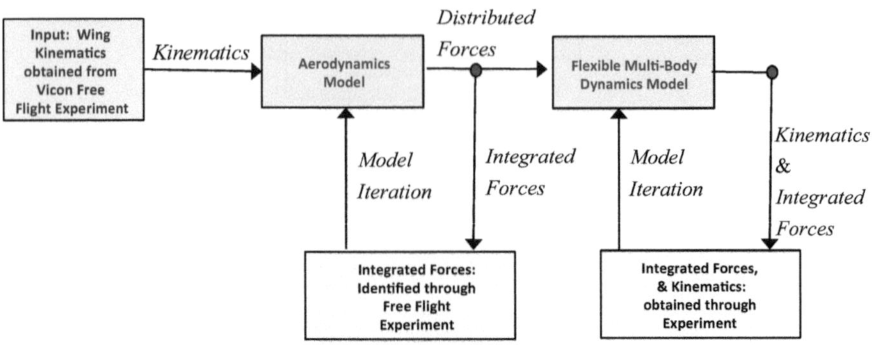

Figure 2.14 Modeling methodology: workflow stage 5.

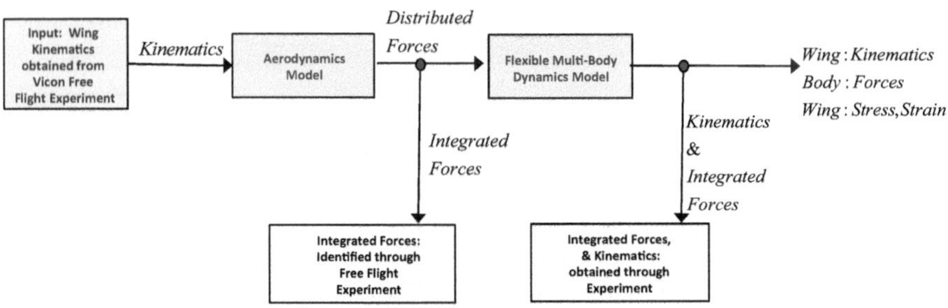

Figure 2.15 Modeling methodology: workflow stage 6.

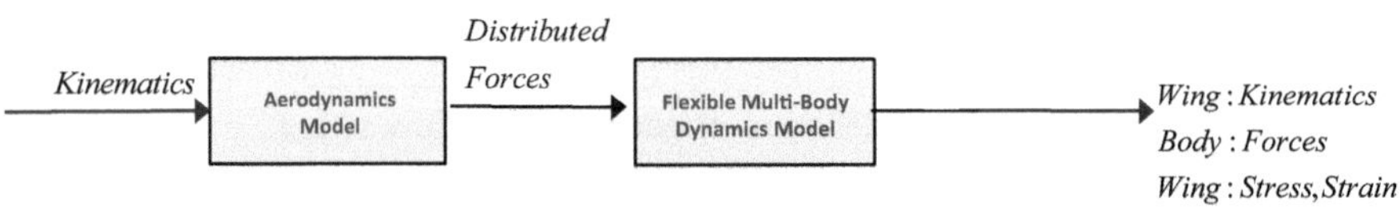

Figure 2.16 Modeling methodology: workflow stage 7.

During the first stage of the development methodology according to (Fig. 2.10) the wing kinematics obtained in the free-flight experiment serve as input for the aerodynamic model. The output of aerodynamic model within its modeling assumptions results in TV and spatially distributed forces on the wing.

Integrated forces are compared with the integrated forces obtained through free-flight test data in the second stage of the model development. Aerodynamic modeling assumptions are adjusted to correlate system ID model and aerodynamic model results. The aerodynamic model and flexible MBD model are verified independently before they are interfaced into a comprehensive model.

Therefore, in the third step of the model development, flexible MBD model output is verified by employing inertial forces and wing kinematics obtained though the vacuum chamber experiment. Leading edge spar kinematics serve as input variable (Fig. 2.12).

In the next step of the model development, aerodynamic and flexible MBD models are combined. The distributed forces resulting from the aerodynamics model are applied on the structural representation of the wing in the flexible MBD model (Fig. 2.13).

The output kinematics and integrated forces are verified against the experiment in the following stage. The model is iterated to archive correlation to experiments in workflow stage 5 (Fig. 2.14).

The output of the iterated flexible MBD model considering aeroelasticity is wing kinematics body forces and wing stress and strain (Fig. 2.15). The overall model operates with wing kinematics obtained from Vicon Free Flight experiment at this stage.

At the final stage of the model development, the input kinematics is removed and the model is coupled. Output wing kinematics serves as input kinematics. The model is initialized by experimental kinematics (Fig. 2.16).

The configuration of the ornithopter model can now be changed and the coupled model can now serve for design optimization purposes.

In summary, wing kinematics obtained from experiment serves as input variable for model development. Individual models are correlated against experiment, integrated into a comprehensive model, and further correlated. Experimental kinematics is removed and the resulting model is coupled to archive a model to change configurations. This model is initialized by experimental kinematics.

REFERENCES

[1] O. Bauchau, Flexible Multi Body Dynamics, Springer, Atlanta, Georgia, 2010, http://dx.doi.org/10.1007/978-94-007-0335-3.

[2] A.A. Shabana, Flexible multibody dynamics: review of past and recent developments, Multibody System Dynamics 1 (1997) 189–222.

[3] J. Grauer, J.E. Hubbard, Modeling and System Identification of an Ornithopter Flight Dynamics Model, 2011.

[4] K. Panagoitis, Model Order Reduction in Structural Mechanics, Coupling the Rigid and Elastic Multi Body Dynamics, Professur für Fahrzeugmodellierung und –Simulation, Institut fuer Bahnfarzeuge und Bahntechnik, Technische Universitaet Dresden, 2009.

[5] L. Meirovitch, Fundamentals of Vibrations, McGraw-Hill Higher Education, New York, 2001.

[6] MacNeal-Schwendler Corporation (MSC), Motion Workspace Guide Introduction – Simxpert, MSC, Newport Beach, CA, 2011.

[7] B. Oliver, DYMORE: A Finite Element Based Tool for the Analysis of Nonlinear Flexible Multibody Systems, Georgia Institute for Technology, Atlanta, 2009.
[8] Johnson Aeronautics, CAMRAD II, Comprehensive Analytical Model of Rotorcraft Aerodynamics and Dynamics, Analytical Methods, Inc., 2010.
[9] Comsol, Comsol Multipurpose Software Platform, 2014 [Online]. Available: www.comsol.com.
[10] SimXpert, Integrated Multidisciplinary Enterprise Simulation Solution for the CAE Analyst, February 2012 [Online]. Available: http://www.mscsoftware.com/Products/CAE-Tools/simxpert.aspx.
[11] J.-S. Lee, J.-K. Kim, D.-K. Kim, J.-H. Han, Longitudinal flight dynamics of bio-inspired ornithopter considering fluid-structure interaction, Journal of Cuidance, Control, and Dynamics 34 (3) (May–June 2011) 667.
[12] J.P. Dyhr, N.J. Cowan, D.J. Colmenares, K.A. Morgansen, T.L. Daniel, Autostabilizing airframe articulation: animal inspired air vehicle control, in: 51st IEEE Conference on Decision and Control, March 2012.
[13] I. Faruque, J. Humbert, Dipteran insect flight dynamics: Part I – longitudinal motion about hover, Journal of Theoretical Biology (2010) 1–15.
[14] G.K. Taylor, R. Zbikowski, Nonlinear time-periodic models of the longitudinal flight dynamics of desert locusts *Schistocerca gregaria*, Journal of Experimental Biology (2003) 2803–2829.
[15] R.C. Nelson, Flight Stability and Automatic Control, second ed., WCB McGraw-Hill, 1998.
[16] W. Shyy, H. Aono, S.K. Chimakurthi, P. Trizila, C.K. Kang, C.E. Cesnik, H. Liu, Recent progress in flapping wing aerodynamics and aeroelasticity, Progress in Aerospace Sciences (2010), http://dx.doi.org/10.1016/j.paerosci.2010.01.001.
[17] K. Sibilski, Dynamics of micro-air-vehicle with flapping wings, Acta Polytechnica 44 (2) (2004) 15–22.
[18] C. Orlowski, Flapping Wing Micro Air Vehicles: An Analysis of the Importance of the Mass of the Wings to Flight Dynamics, Stability, and Control (Ph.D. thesis), University of Michigan, 2011.
[19] X. Deng, L. Schenato, W. Chung Wu, S. Sastry, Flapping flight for biomimetic robotic insects: part I – system modeling, IEEE Transactions on Robotics 22 (4) (August 2006) 776–788.
[20] X. Deng, L. Schenato, S. Sastry, Flapping flight for biomimetic robotic insects: part II – flight control design, IEEE Transactions on Robotics 22 (4) (August 2006) 789–803.
[21] I. Faruque, Control-Oriented Reduced Order Modeling Dipteran Flapping Flight (Ph.D. thesis), University of Maryland, 2011.
[22] J. Lee, J. Kim, J. Han, Cycle-Averaged Mean and Zero-Mean Time-Varying Flight Dynamics of the Ornithopters at Trim, 2010.
[23] J. Grauer, J. Hubbard, Modeling of ornithopter flight dynamics for state estimation and control, in: IEEE, American Controls Conference, June 2010.
[24] M.R. Waszak, L.N. Jenkins, P. Ifju, Stability and control properties of an aeroelastic fixed wing micro aerial vehicle, vol. 4005, AIAA Flight Mechanics, Montreal, 2001, p. A01-37226.
[25] M.R. Waszak, C.S. Buttrill, D.K. Schmidt, NASA Technical Memoradum, Modeling and Model Simplification of Aeroelastic Vehicles: An Overview, vol. 107691, September 1992.
[26] S.S. Gheorghe Bunget, in: The 15th International Symposium on: Smart Structures and Materials & Nondestructive Evaluation and Health Monitoring, March 2008, 69282F-69282F-12.

[27] J.M. Dietl, E. Garcia, Stability in ornithopter longitudinal flight dynamics, Journal of Guidance, Control, and Dynamics 31 (4) (August 2008) 1157–1162.
[28] D.-K. Kim, J.-S. Lee, J.-H. Han, Improved aerodynamic model for efficient analysis of flapping-wing flight, AIAA Journal 49 (4) (April 2011) 868–872.
[29] M. Bolender, Rigid multi-body equations-of-motion for flapping wing MAVs using kane's equations, in: AIAA, Guidance, Navigation, and Control Conference, August 2009.
[30] A.T. Pfeiffer, J.-S. Lee, J.-H. Han, H. Baier, Ornithopter flight simulation based on flexible multi-body dynamics, Journal of Bionic Engineering 7 (2010).
[31] T. Rashid, The Flight Dynamics of a Full-Scale Ornithopter (Master's thesis), 1995.
[32] W. Shyy, Y. Lian, J. Tang, D. Viieru, H. Liu, Aerodynamics of Low Reynolds Number Flyers, Cambridge University Press, 2008.
[33] P. Gopalakrishanan, Unsteady Aerodynamic and Aeroelastic Analysis of Flapping Flight (Dissertation), Virginia Polytechnic Institute and State University, VA, 2008.
[34] S. Swartz, Skin and Bones: The Mechanical Properties of Bat Wing Tissue, 2010, pp. 109–126.
[35] J.D. DeLaurier, An aerodynamic model for flapping-wing flight, The Aeronautical Journal 97 (964) (April 1993) 125–130.
[36] R.P. Malhan, Investigation of Aerodynamics of Flapping Wings for Micro Air Vehicle Applications (Dissertation), 2013.
[37] M. Okamoto, A. Azuma, Experimental study on aerodynamic characteristics of unsteady wings at low Reynolds number, AIAA Journal 43 (12) (2005) 2526–2536.
[38] R.L. Harmon, J.E. Hubbard, Aerodynamic Modeling of a Flapping Membrane Wing Using Motion Tracking Experiments (Master's thesis), 2008.
[39] W.P. Walker, Unsteady Aerodynamics of Deformable Thin Airfoils, Virginia Polytechnic Institute and State University, 2009.
[40] B. Roget, J. Sitaraman, R. Harmon, J. Grauer, J. Hubbard, S. Humbert, A computational study of flexible wing ornithopter flight, in: 26th AIAA Applied Aerodynamics Conference, August 18–21, 2008.
[41] A. Wissa, N. Guerreiro, J.A. Grauer, J. Hubbard, M. Frecker, C. Altenbuchner, Y. Tummala, R. Roberts, Flight testing of novel compliant spines for passive wing morphing on ornithopters, in: Structural Dynamics, and Materials and Co-located Conferences, April 2013.
[42] C. Altenbuchner, J.E. Hubbard, A.A. Wissa, Free flight validation of a flexible-multibody structural dynamics model of a bioinspired ornithopter, in: 51st AIAA Aerospace Sciences Meeting, January 2013.
[43] B.K. Stanford, P.S. Beran, Analytical sensitivity analysis of an unsteady vortex-lattice method for flapping-wing optimization, Journal of Aircraft 47 (2) (March–April 2010) 647–662.
[44] A. Hart, L. Ukeiley, Low Reynolds number unsteady aerodynamics over a pitching-plunging flat plate, in: 48th AIAA Aerospace Sciences Meeting Including the New Horizons Forum and Aerospace Exposition, January 4–1, 2010.
[45] M.R. Visbal, High-fidelity simulation of transitional flows past a plunging airfoil, AIAA Journal 47 (11) (November 2009) 2685–2697.
[46] W. Yuan, L. R, H. E, K. M, Numerical and experimental simulations of flapping wings, International Journal of Micro Air Vehicles (September 2010) 181–209.
[47] I.H. Tuncer, M.F. Platzer, Thrust generation due to airfoil flapping, AIAA Journal, 34 (February 1996) 324–331.
[48] I. Tuncer, R. Walz, M. Platzer, A Computational Study of the Dynamic Stall of a Flapping Airfoil, AIAA Paper 98-2519, June 1998.

[49] M. Yu, Z.J. Wang, H. Hu, High-fidelity flapping-wing aerodynamics simulations with a dynamic grid spectral difference method, in: ICCFD7 Seventh International Conference on Computational Fluid Dynamics, July 9–13, 2012.
[50] Dae-Kwan Kim, J.-S. Lee, J.-H. Han, Improved aerodynamic model for efficient analysis of flapping-wing flight, AIAA Technical Journal 49 (4) (April 2011).
[51] U. Pesavento, Z.J. Wang, Flapping wing flight can save aerodynamic power compared to steady flight, Physical Review letters PRL 103 (118102) (2009).

CHAPTER THREE

Bioinspired Flapping Wing Test Platform Used to Implement Modern Modeling Methodology

3.1 DETAILS OF THE TEST PLATFORM

The primary test platform in this work is the Morpheus Lab custom-built test ornithopter (ML101) with a solid leading edge spar (Fig. 3.1). This configuration serves as the baseline test platform for the model development.

The secondary test platform is a modified version of a commercially available ornithopter designed by Shawn Kinkade (MSK004). See Fig. 3.2.

The experimental ornithopters have spatially distributed stiffening carbon fiber spars (Fig. 3.4) covered by a thin membrane of rip stop material (Fig. 3.3).

Stiffening finger spars, spar one to four, a leading edge spar, and a diagonal spar are located on the wing (Fig. 3.5) and provide structural integrity for the wing.

Such a structure provides full elastic behavior while undergoing large deflections such as the flapping motion. The wings of this test platform are divided into two essential zones: the luff region and the thrust flap region, shown in Fig. 3.3. During a wing stroke, the camber of the luff

Figure 3.1 Primary test platform Morpheus Lab custom-build test ornithopter (ML101).

Modern Flexible Multi-Body Dynamics Modeling Methodology for Flapping Wing Vehicles
ISBN: 978-0-12-814136-6
http://dx.doi.org/10.1016/B978-0-12-814136-6.00003-2

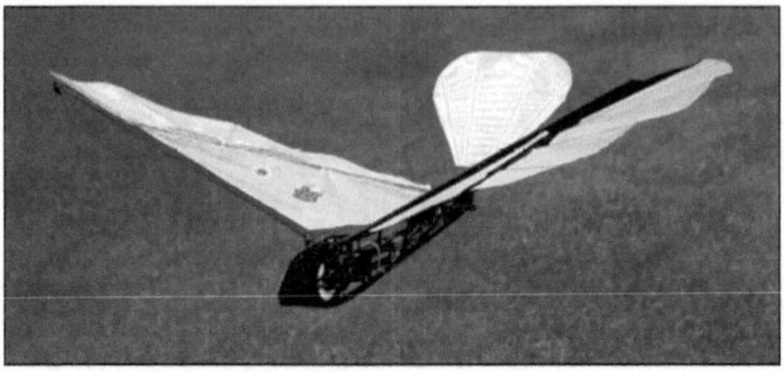

Figure 3.2 Secondary test platform modified ornithopter Shawn Kinkade (MSK004).

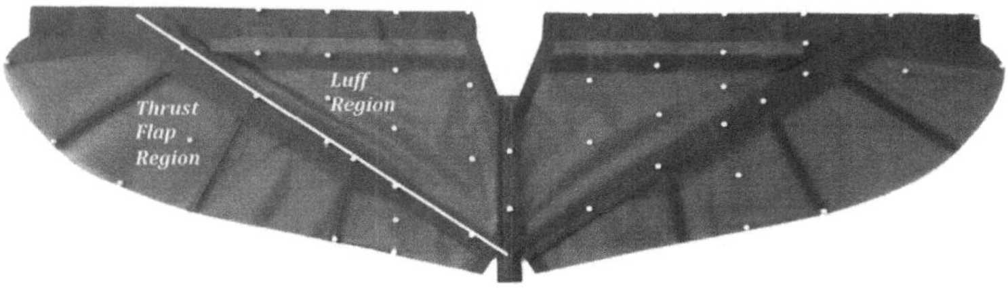

Figure 3.3 Wing structure bioinspired ornithopter test platform (ML101).

Figure 3.4 Design feature wing spars (ML101); (A) leading edge spar, (B) diagonal spar, (C) finger spar (ML101).

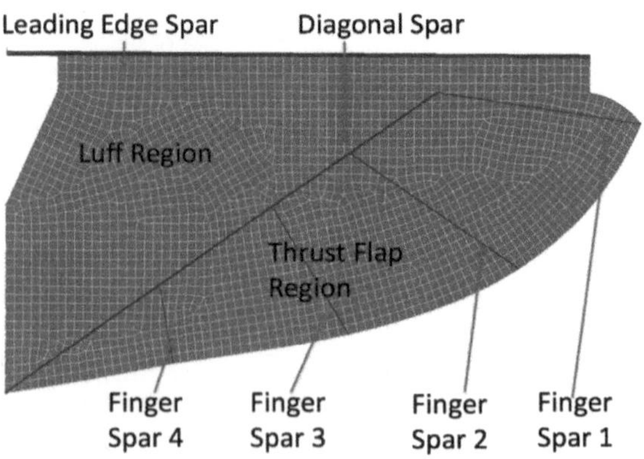

Figure 3.5 Schematic and nomenclature stiffening carbon fiber spar configuration.

region forms an airfoil shape camber, while the thrust flap region bends out and forms a self-activated flap. These combined characteristics found in birds and bats are observed in the bioinspired research vehicles. The ornithopter test platforms consist of a wing membrane (Fig. 3.3) with stiffening carbon fiber spars, a simplified architecture found in bats (Fig. 3.6). In both the vehicle and bats, a camber forms in the wing membrane and forms out during a wing stroke. With camber forming in combination with a complex structure of joint motions, a positive and negative angle of attack (AoA) shapes out during a wing beat. The mechanisms utilized in these bioinspired test platforms are analogous to bats.

Like a bird, the test platform is outfitted with a self-activated flap region: the so-called "thrust flap region" (Fig. 3.3). According to bird flight, flexible convert feathers (Fig. 3.7) act like self-activated flaps similar to the thrust flap region [1].

The primary test platform (Fig. 3.1) features the wing (Fig. 3.3), the tail (Fig. 3.8), the fuselage (Fig. 3.9), and the stiffening carbon fiber spars (Fig. 3.4), which make up the leading edge spar, the diagonal spar, and, the finger spars. Table 3.1 specifies the material and dimensions of these primary test platform features.

The test platforms are radio-controlled, and the flapping motion is driven by an electric motor through a gear box and a bar linkage. The resulting

Figure 3.6 A bat (*Cynopterus brachyotis*) in flight. *Image National Park Service courtesy of Christopher Iezzoni; W. Shyy, Y. Lian, J. Tang, D. Viieru, H. Liu, Aerodynamics of Low Reynolds Number Flyers, Cambridge University Press, New York, NY, 2008.*

Figure 3.7 Lower wing surface of a natural avian flyer photo. *Image: National Park Service courtesy of Christopher Iezzoni indicating flap region in bird flight.*

Figure 3.8 Design feature tail (ML101).

Figure 3.9 Design feature fuselage (ML101).

Table 3.1 Material and dimensions of ornithopter structures

Element	Thickness diameter (mm)	Material	Density (kg/m^3)
Wing membrane	0.06	Polyester ripstop	500
Leading edge spar	3.6	Carbon fiber	1750
Diagonal spar	1.8	Carbon fiber	1750
Finger spar 1–4	1	Carbon fiber	1750

motion drives the leading edge spar close to the fuselage where the bar linkage is attached to a shoulder joint. This provides further details about the test platform [2,3].

The secondary test platform, MSK004, has an analogous configuration to the primary test platform (ML101), but differs according to aircraft specification found in Table 3.2. The main difference of the secondary test platform Fig. 3.34 lies in the scale. MSK004 is a larger scale.

The flapping angle (FA) and the upstroke/downstroke sequence for the test platform are defined next and were used throughout this work [4]. At

Table 3.2 Aircraft specifications

Ornithopter platform	Symbol	ML101 (Harmon and Hubbard [4]; Wissa et al. [7])	MSK004 (Harmon and Hubbard [4]; Grauer and Hubbard [5])
Wing span	b_w	1.07 m	1.21 m
Flapping rate	f_r	4–6 Hz	2.5–5 Hz
Total Mass	m_t	0.42 kg	0.45 kg
Forward speed	v_f	2.5–8.5 m/s	2.5–8.5 m/s
Wing mean aerodynamic chord	$\bar{c}_w$	0.22 m	0.29 m
Wing area	S_w	0.23 m^2	0.3 m^2
Wing aspect ratio	AR_w	4.97	4.40
Tail span	b_t		0.20 m
Tail mean aerodynamic chord	$\bar{c}_t$		0.20 m
Tail area	S_t		0.04 m^2
Tail aspect ratio	AR_t		1.50
Reynolds number regime	Re	1.9×10^4 and 23×10^4	1.9×10^4 and 23×10^4
Average Re number for typical	Re_t	10×10^4	13×10^4

maximum negative FA, the FA for an extended wing is defined at zero. The upstroke to downstroke to transition occurs at the maximum positive AoA. The downstroke to upstroke transition occurs at the minimal negative AoA. Fig. 3.10 shows the upstroke/downstroke transition.

Fig. 3.11 shows the primary (ML101) scale test platform undergoing a flapping cycle and the upstroke/downstroke sequence.

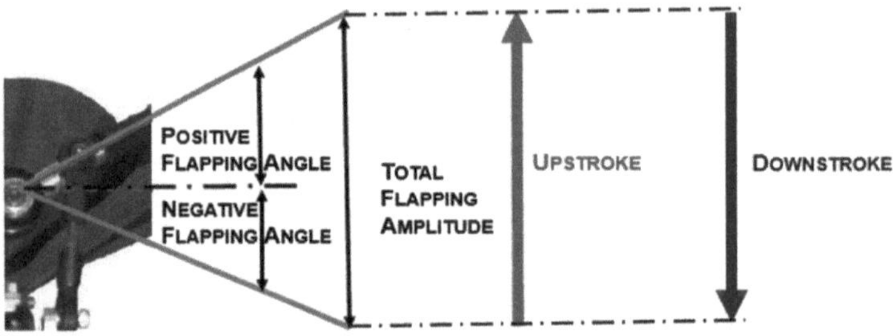

Figure 3.10 Definition flapping angle and upstroke and downstroke on shoulder joint/bar linkage (ML101) [4].

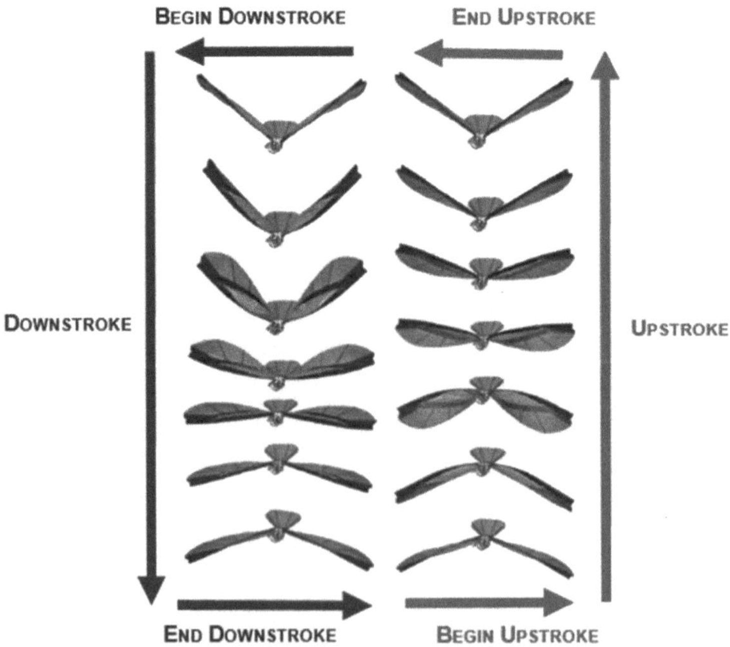

Figure 3.11 Upstroke and downstroke sequence ornithopter (ML101) scale. *Image credit R.L. Harmon, J.E. Hubbard, Aerodynamic Modeling of a Flapping Membrane Wing Using Motion Tracking Experiments (Master's Thesis), 2008.*

In summary, this section described the design features and specifications for the primary (ML101) and secondary test platforms (MSK004) used in this work for modeling and verification purposes.

3.2 EXPERIMENTAL DATA SETS OF BIOINSPIRED FLAPING WING ROBOTIC SYSTEM FOR MODEL VERIFICATION

In the following sections, the ornithopter experiments utilizing the ML101 and MSK004 test platforms, and resulting data used in this work are reviewed. First, a brief description regarding the archival ornithopter tests used for verification of this model is provided. In addition, the data extracted and used in the frame of this work are presented. The following is a description of the free-flight test and vacuum chamber experiment developed to verify and validate the assumptions and constraints of the model methodology presented in this body of work. The first archival experiments are named clamped test E1 and system-ID experiment E1-I. Further experiments include a free-flight experiment, denoted as free-flight test E2 and the vacuum chamber experiment, denoted as E3 in the following work.

3.2.1 The Clamped Test Experiment—E1

Experimental flight test data are instrumental to model the complex problem given. The University of Maryland has established a unique experimental flight test database of its ornithopter fleet that contains wing surface motions and kinematics, as well as force time histories collected using a Vicon vision data collection system. Wing kinematics and integrated forces on the ornithopter fuselage during flapping were obtained for the ML101 and MSK004 test platforms. Flapping frequencies of 5 and 6.17 Hz were obtained for test platform ML101 and 4.5 Hz for test platform MSK004. In E1, the Vicon Vision system uses six high-speed cameras at 350 frames/s to track the 3D position of 3-mm reflective markers attached to the wing surface as seen in Fig. 3.12.

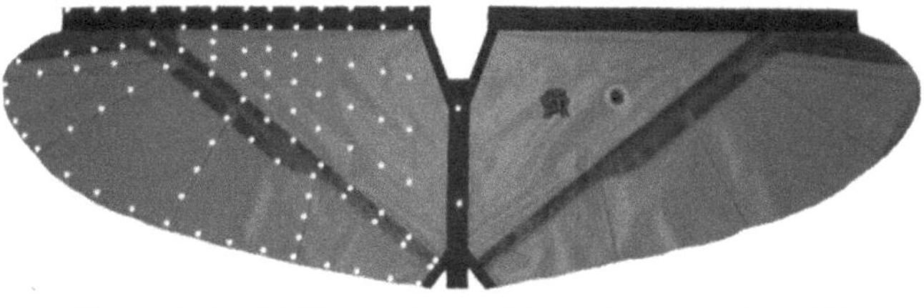

Figure 3.12 MSK004 wings with tracking markers [4].

The ornithopter frame was rigidly attached to a ground stand. The vertical propulsive force (VPF) and the horizontal propulsive force (HPF) produced by the flapping motion were measured with a six degrees of freedom (DoFs) load cell mounted to the ground stand that supports the ornithopter. Harmon [4] provides additional details. The complete experimental setup is shown in Fig. 3.13.

Note that the clamped condition provided in load cell testing disturbs the overall system dynamics and kinematics because it interferes with the pitch moments and flight aerodynamics. Constraints are on rotational and translational DoFs of the fuselage body-fixed reference frame as defined in Chapter 4.

Constraint integrated forces and moments measured with the six DoFs load cell were used for the verification of overall integrated force production of the model simulated in clamped conditions. Wing kinematics used for a proof-of-concept study of the aerodynamic model development was amended for the clamped conditions by an adjustment of circulatory effect as described in Chapter 5. Data resulting from E-1 extracted for this work are presented next.

3.2.1.1 Results of Experiment and Integrated Forces—E1

The total measured horizontal and vertical propulsive force (HPF/VPF) with the six degree of load cell, by Harmon [4], was used from the Morpheus Lab database and utilized for the model verification.

HPF and VPF data versus the stroke angle obtained for the ML101 test platform, at flapping frequencies of 5.0 and 6.17 Hz, were extracted and

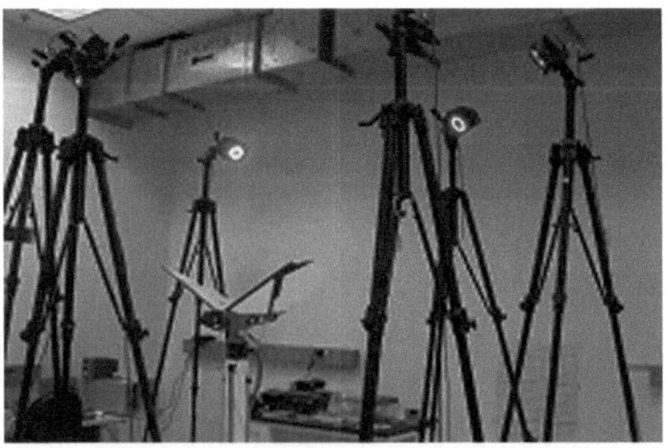

Figure 3.13 MSK004 and Vicon camera system for the measurement of ornithopter wing and configuration motions [4].

processed from the previously mentioned database according to Figs. 3.14–3.17.

Mean absolute vertical propulsive force and mean horizontal propulsive force (MHPF) versus flapping frequencies were extracted from the E1 experiment database and used for model verification. Error bars were

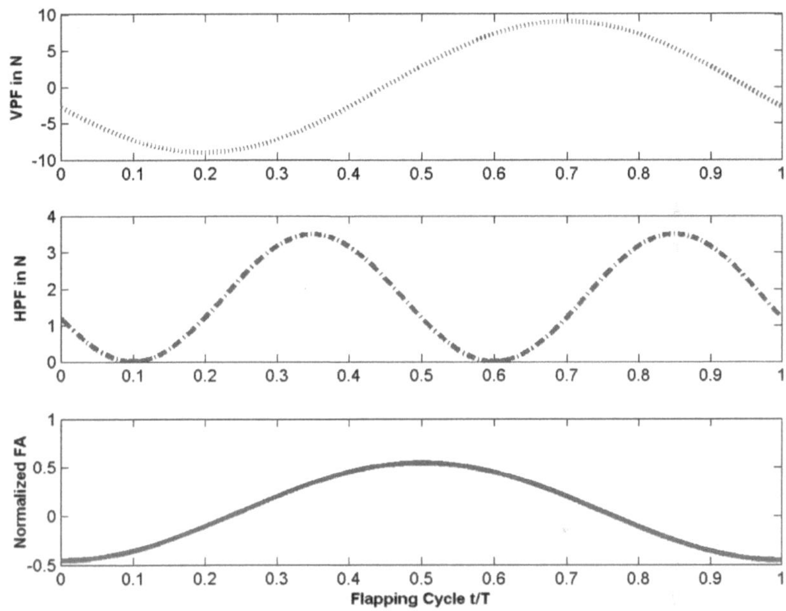

Figure 3.14 Harmon aerodynamic model results and bench test results measured integrated forces ML101 at 5 Hz flapping frequency, over a flapping cycle t/T, (A) vertical propulsive force (VPF), (B) horizontal propulsive force (HPF), (C) normalized FA.

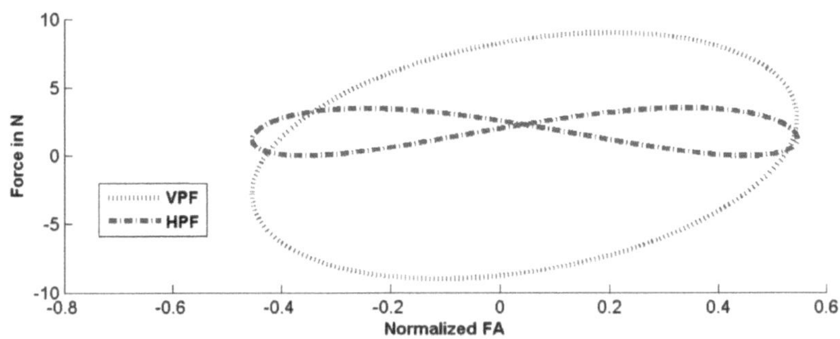

Figure 3.15 Harmon bench test results measured integrated forces ML101 at 5 Hz flapping frequency, over a flapping cycle t/T, (A) vertical propulsive force (VPF), (B) horizontal propulsive force (HPF) versus normalized flapping angle (FA).

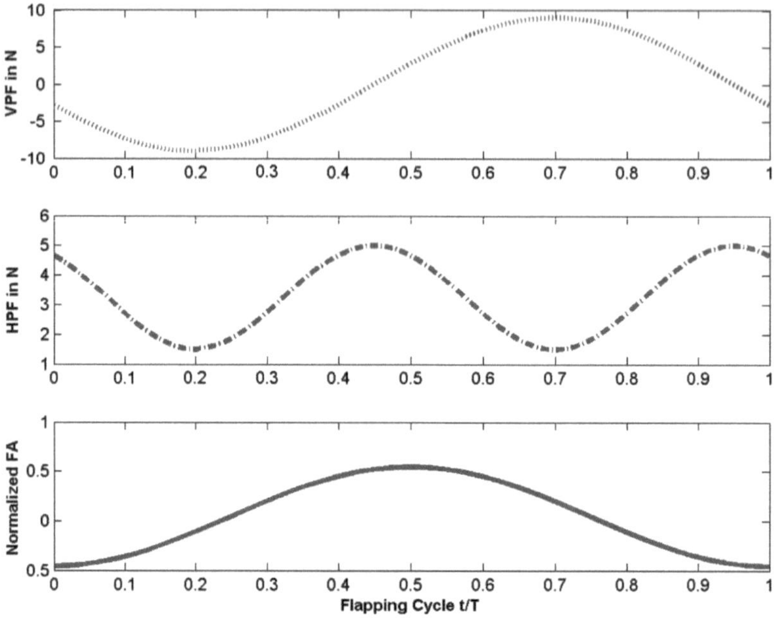

Figure 3.16 Harmon aerodynamic model results and bench test results measured integrated forces ML101 at 6.17 Hz flapping frequency, over a flapping cycle t/T, (A) vertical propulsive force (VPF), (B) horizontal propulsive force (HPF), (C) normalized flapping angle (FA).

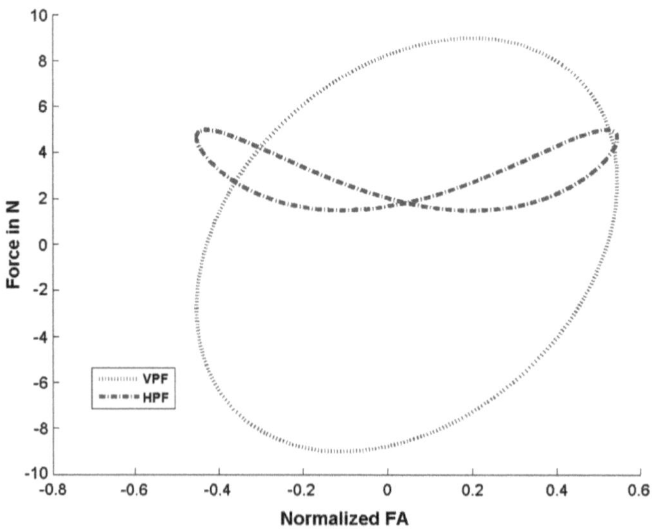

Figure 3.17 Harmon bench test results measured integrated forces ML101 at 6.17 Hz flapping frequency, over a flapping cycle t/T, (A) vertical propulsive force (VPF), (B) horizontal propulsive force versus normalized flapping angle (FA).

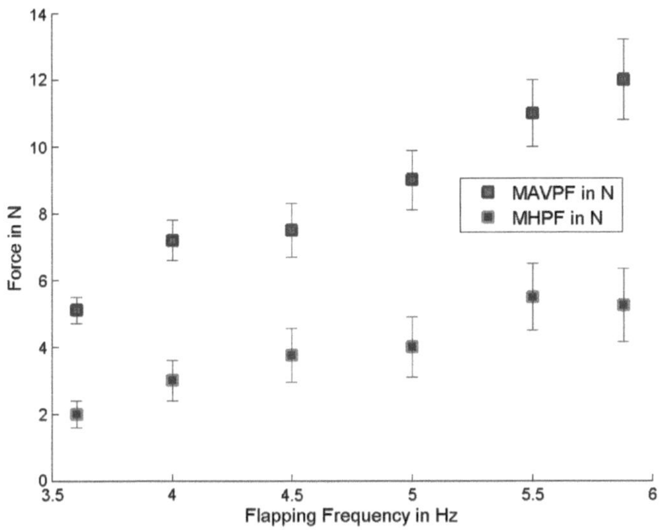

Figure 3.18 Results and bench test results measured integrated forces ML101 versus flapping frequency, over a flapping cycle t/T, (A) mean absolute value vertical propulsive force (MAVPF), (B) mean horizontal propulsive force (MHPF) versus flapping angle [4].

included according to the properties of load cell and the data measurement system used (Fig. 3.18).

3.2.1.2 Wing Kinematics—E-1

ML101 wing kinematics obtained in E-1 used as input for aerodynamic model A are described in Section 5.2. A data set of 110 tracking markers, at a flapping frequency of 6.17 Hz, was used for kinematic simulation verification and input to aerodynamic model A. Data were measured at a resolution of 1−2 mm, at a rate of 350 Hz. A data array, for one full flapping cycle of 56 position states for 110 tracking markers, in a fuselage fixed reference frame, was used and extracted from the database. Fig. 3.19 shows the three position states of used data.

3.2.2 The System-ID Experiment—E1-I

System identification of the ornithopter wing kinematics and development of the aerodynamics model using free-flight data was accomplished using the MSK004 test platform [5,6]. Aerodynamic model results obtained from Grauer and Hubbard[5] were used to verify the resulting aerodynamics model in this work. In the previously mentioned experiments, eight cameras were placed to capture a flight volume of 10-m long by 6-m wide, and 5-m

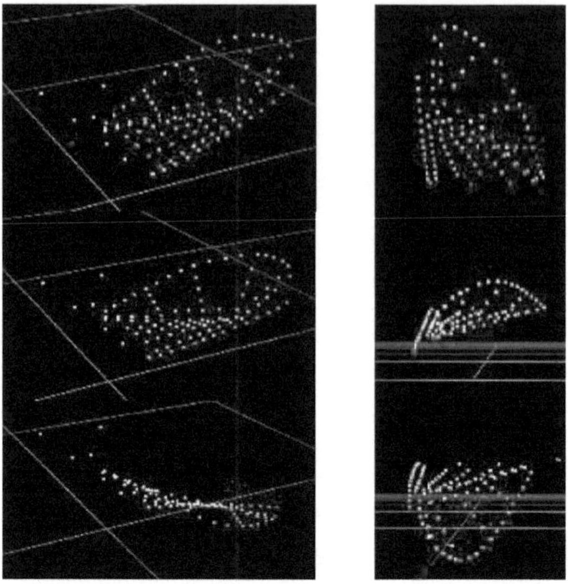

Figure 3.19 Vicon Vision wing kinematic sequence—three position state tracking markers $i = 1-110$, (A) isometric view, (B) side view in a fuselage fixed reference frame CB0.

tall [5]. At 500 Hz, these cameras recorded the spatial position of the retroreflective markers placed on the ornithopter's fuselage, wings, and tail (Fig. 3.20) as the ornithopter was flown through the control volume. This method and the amount of reflective markers were sufficient to perform a system identification of a three body-rigid body dynamics model and aerodynamics model in straight and level flight. Resulting aerodynamic model results are based on free flight and provide an appropriate source for

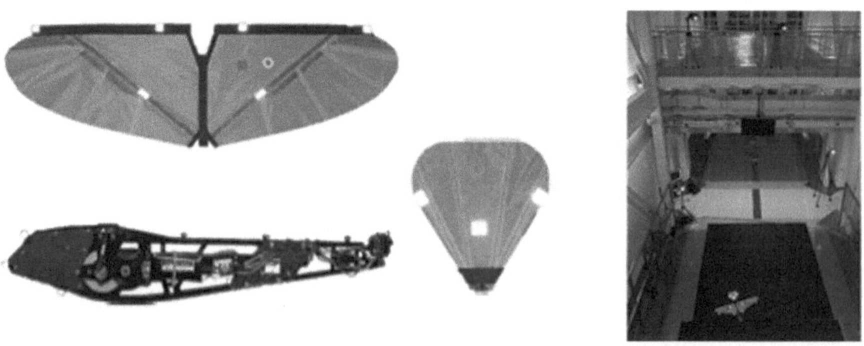

Figure 3.20 Vicon camera system for the measurement of ornithopter wing and configuration motions [5].

verification of free-flight aerodynamics models B and C (see Chapter 5 Aerodynamics). The Aerodynamics model resulting from experiment E1-I by Grauer provides integrated forces acting on the fuselage for MSK004 ornithopter configuration. Models B and C provide spatially distributed aerodynamic forces over the wings, and integrated first to be compared with E1-I. Grauer and Hubbard [5,6] provide reference details of the tests and resulting aerodynamics model.

3.2.2.1 Integrated Aerodynamic Force—E1-I

Following the data from the system ID aerodynamic model, which provided integrated forces and moments in free flight for ornithopter configuration MSK004, was used for verification of the aerodynamics model resulting in this work. Vertical and horizontal propulsive forces during a wing stroke were computed using Grauer's model and plotted in Fig. 3.21. The data serve for verification purposes.

3.2.3 Free-Flight Experiment—E2

A free-flight test exposing flexible wing kinematics was designed in the framework of this work and performed by Wissa et al. [7]. The flight test was

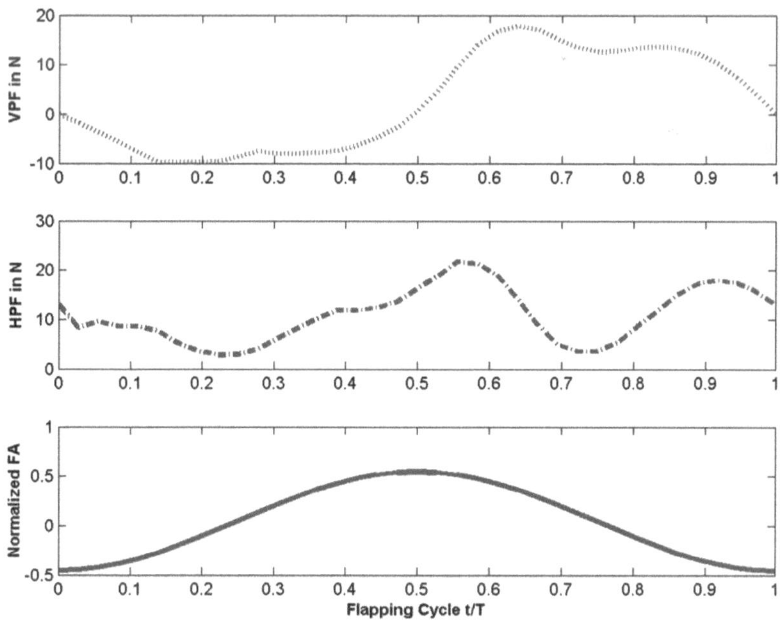

Figure 3.21 Aerodynamic forces on the wings—results system ID model by Grauer [5]. *FA*, flapping angle; *HPF*, horizontal propulsive force; *VPF*, vertical propulsive force.

conducted with the assumptions and the constraints of this work. The experiment E-2 was designed to capture wing motion profile for verification of the flexible multi-body dynamics model. This resulted in the requirement of a Vicon Vision software capable of exposing wing flexibility in contrast to the commonly used Vicon Vision systems, which fit rigid body's in-between reflective markers. This requirement resulted in the flight test to be conducted and facilitated at the Wright-Patterson Air Force Base (WPAFB) in the Air Force Research Laboratory's (AFRL) indoor small Unmanned Aerial Systems (sUAS) laboratory. The facility hosts the largest Vicon Vision motion capture system lab in the United States, and the test room is fitted with 60 motion cameras tracking retroreflective marker positions with a precision of 1 mm. The following describes the test setup for the experiment, including wing configuration, marker placement, flight test facility, and test setup.

The kinematics of the test platform ML101 was measured in free flight to verify the flexible multi-body dynamics model. For the test, retroreflective markers were placed on the wing (Fig. 3.3). A diameter of 6.35 mm of the half sphere retroreflective markers was chosen.

The chosen marker size lies on the larger spectrum of available markers but ensured the likelihood of full visibility and minimized the likelihood of marker dropout. Precautions were necessary since the desired wing kinematics resulted in the highest marker refinement and the most number of markers flown in the test facility and ensured results. The markers were asymmetrically distributed over both the left and right wings to gain high refinement of symmetric wing kinematics of the flight platform, in addition to balancing the weight distribution on both wings. Compared with a symmetric distribution on one wing, this resulted in the feasibility of more tracking points as the distance between the tracking points, for appropriate motion capturing was provided. Wing tip markers were placed on both wings to test symmetry of wing kinematics during the flight test (Fig. 3.22).

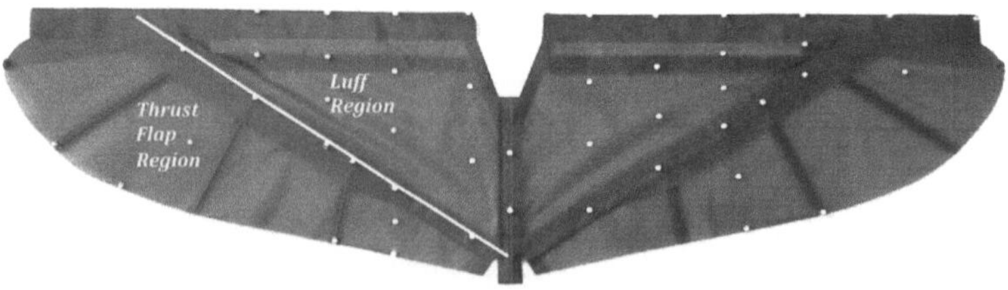

Figure 3.22 Asymmetric retroreflective markers on the wing (seen as white dots) and wing tip location (marked in red).

Forty-four markers were placed on the wing according to Fig. 3.22. Distribution was chosen based on blade element selection for the aerodynamic model. The number of markers on the ML101 test platform was 53 in total. To determine fuselage kinematics, five markers were placed on the ornithopter fuselage. Three were placed on the tail to monitor control inputs, and one was mounted on the wing root to track relative wing displacement. The markers on the flight platform are denoted as $i = 1-53$ (Fig. 3.23).

Data, using the Vicon Vision Nexus software, were captured at a sampling rate of 200 frames per second (200 Hz) . The flight test facility was fitted with 60 Vicon Vision cameras (Fig. 3.24). Fig. 3.25 shows the test setup used.

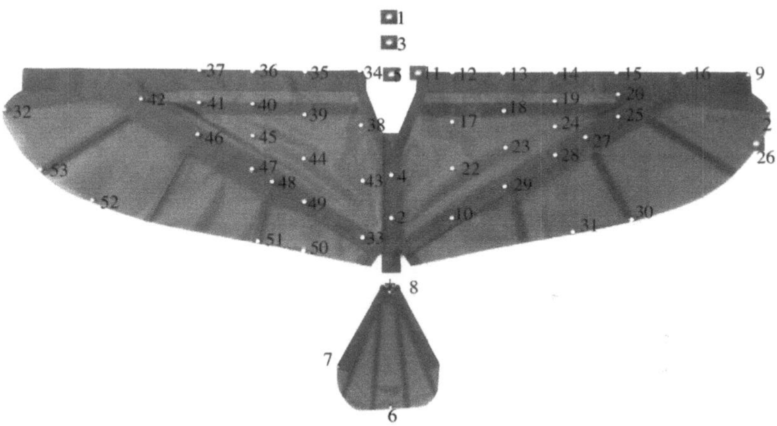

Figure 3.23 Retroreflective marker position on the wing measured with Vicon Vision system markers $i = 1-53$. Markers $i = 1-5$ were mounted on the fuselage.

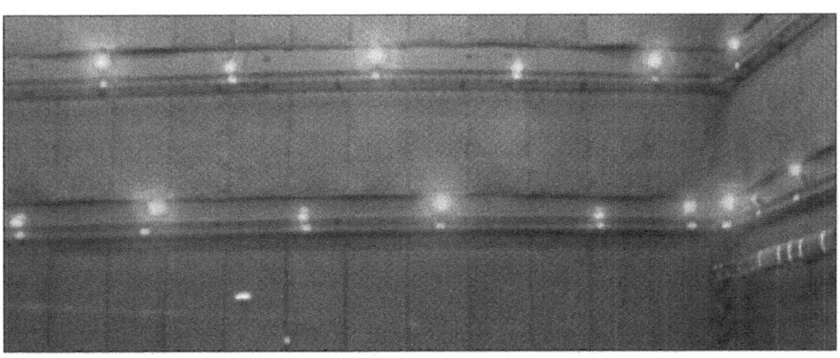

Figure 3.24 Image small Unmanned Aerial System flight test facility—Vicon Vision system cameras [7].

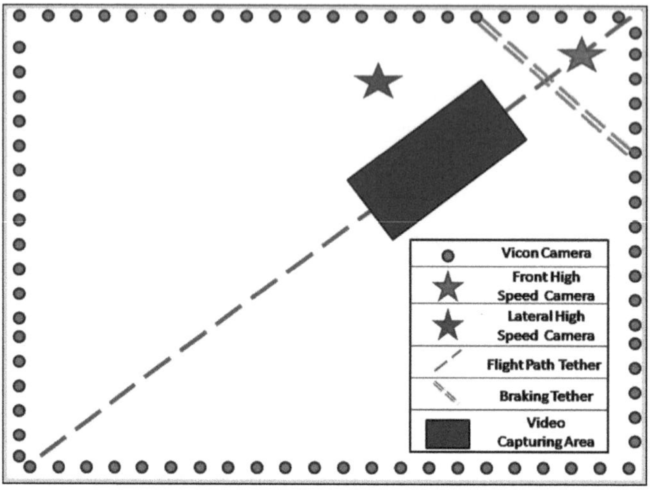

Figure 3.25 Test setup schematic—test chamber dimensions: 70′(W) × 35′(D) × 35′(H) [7].

Additional high-speed cameras captured images of the ornithopter flight. The flight test platform was mounted on a tether to guide flight. Videos were captured using the video capture volume. The test chamber dimensions were 70′(W) × 35′(D) × 35′(H) [7]. The flight platform ML101 was radio-controlled by a ground pilot, and free flight was ensured in the test volume. Results of the free-flight test used for verification of the ornithopter model are described next.

3.2.3.1 Results for Wing Kinematics—E-2

A total of 53 wing kinematics markers were measured in free flight. Data obtained for the verification were captured in the inertia reference frame CI0 (Fig. 3.26). Choice of coordinate notation is according to the standard aircraft dynamics: north, east, down (NED) convention. Fig. 3.26 shows the location of the inertia reference frame CI0 and additional coordinate systems used. For mathematical convenience, the inertia reference frame of the flight test data is rotated 16 degrees into the inertial reference frame CI. The x-axis of CI points in flight direction. R IB is the position vector from the inertial reference frame CI to the body fixed reference frame CB on the fuselage (Fig. 3.26). For verification and modeling purposes, wing position states are rotated and translated in the fuselage fixed reference frame, CB0, located on the wing root. The position vector r B B0 is the translation in-between CB and CB0.

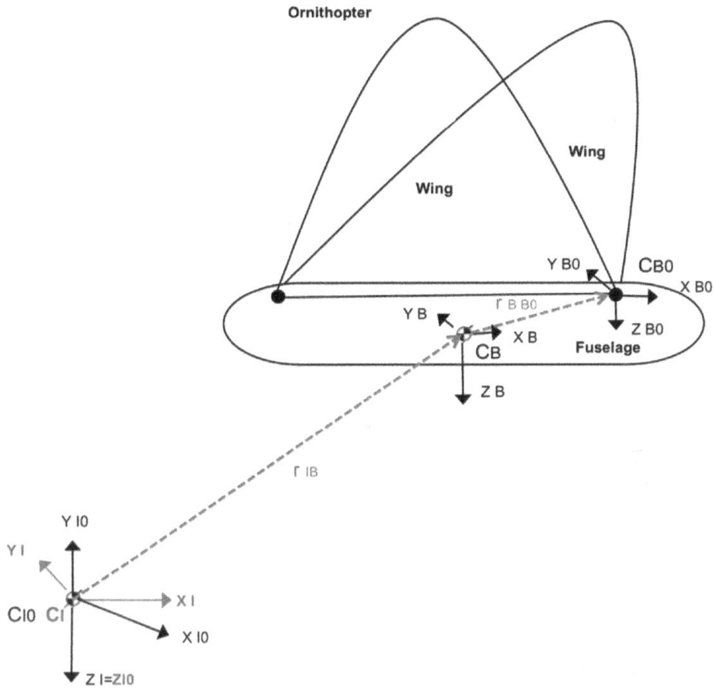

Figure 3.26 Schematic ornithopter and reference frames used for the processing of free-flight test data.

The data matrix used in this work contains 53 marker locations in 125 position states X/Y/Z and were measured in the inertia frame CI0. The data array used for verification and flight test was 34 frames per marker (53), with 3.5 flapping repeatable flapping cycles. The data were captured at a flapping rate of 6.06 Hz, which resulted in 34 data points per marker, per flapping cycle, hence 125 position states for each marker. Free-flight test data for ML101 flight test platform are shown in Figs. 3.27–3.31. Individual lines in Fig. 3.27 represent the position of markers $i = 1$ to 53 during free flight.

Repeatable free-flight flapping cycles were extracted according to the volume shown in Fig. 3.26 and postprocessed using MATLAB.

For clarification, one position state of all tracking markers, $i = 1–53$, is illustrated in Fig. 3.31. The position state is plotted in the fuselage fixed reference frame CB0 and is shown in Fig. 3.26.

Wing kinematics data shown in Figs. 3.27–3.31 were used for modeling and verification purposes throughout this work.

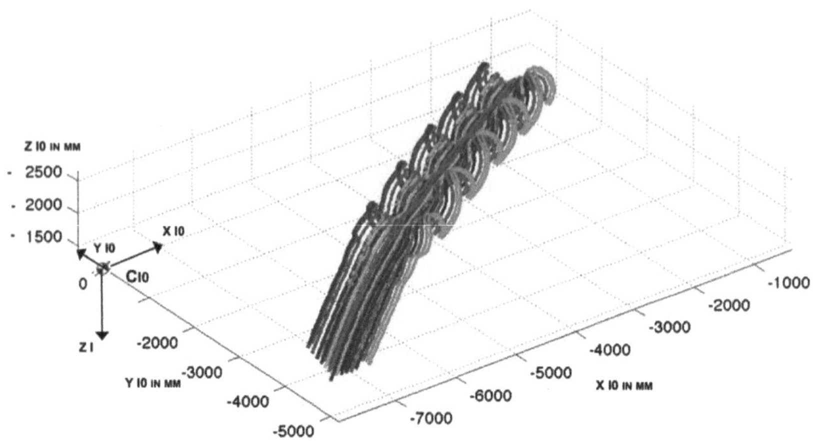

Figure 3.27 Data set marker position during free flight in the inertia reference frame CI0.

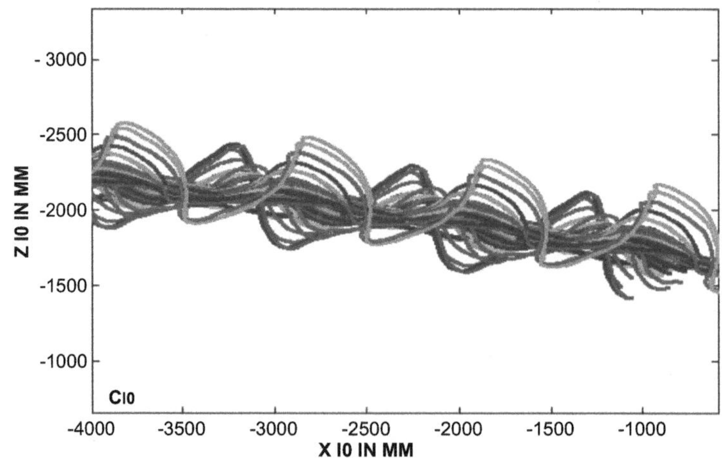

Figure 3.28 Data set marker position $i = 1-53$ during free flight in the inertia reference frame CI0—view ZI0/ZI0.

3.2.3.1.1 Total Forces—E2

The total vertical force (VF) was calculated from flight test data at the fuselage center of mass and origin of CB (Fig. 3.26). Force was obtained from the total vehicle acceleration calculated from the position state of the fuselage fixed-body reference frame, CB, in the inertial reference frame, CI, and the total vehicle mass of test platform ML101. Following VF and horizontal

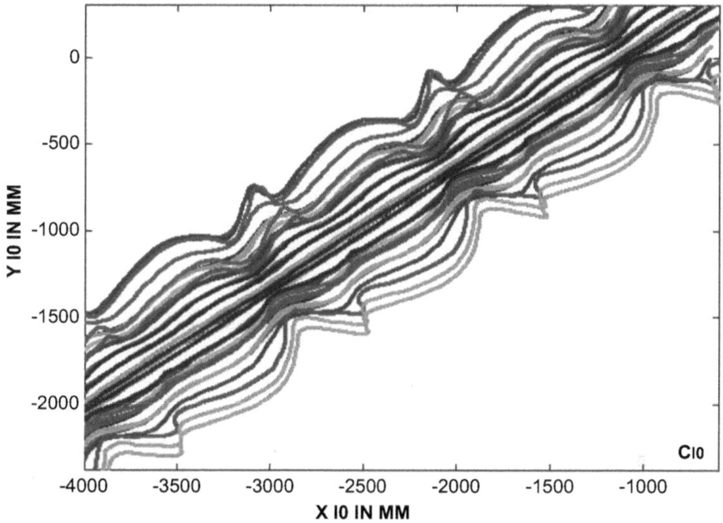

Figure 3.29 Data set marker position $i = 1{-}53$ during free flight in the inertia reference frame CI0—view YI0/XI0.

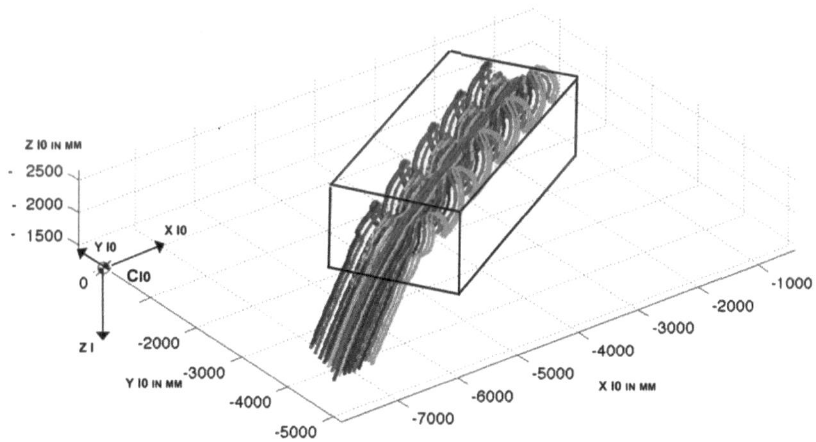

Figure 3.30 Position states all tracking markers on the wing/volume shows test data used for model development—3.5 flapping cycles.

force (HF) (Fig. 3.32), results from the flight test data were used to verify model calculations. VF results from the vehicle acceleration, along the z-axis of CI, and HF, along the x-axis of CI. To determine the state in the flapping cycle, the data were presented in reference to the FA.

In summary, free-flight verification data obtained and used for the modeling development and verification was presented in this section.

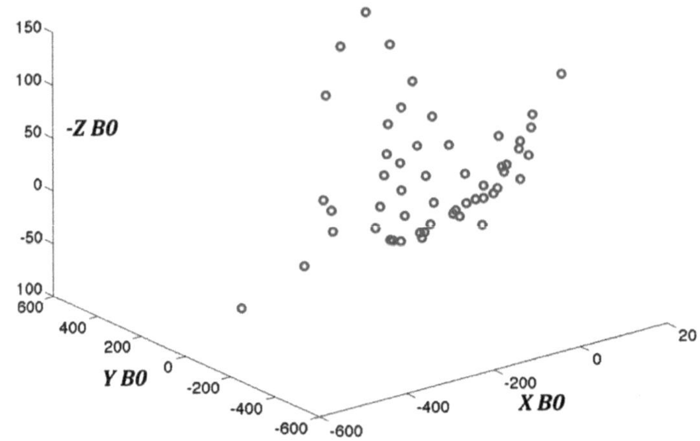

Figure 3.31 Single position state on ornithopter shows all tracking markers $i = 1$ to 53 in the fuselage fixed reference frame CB0.

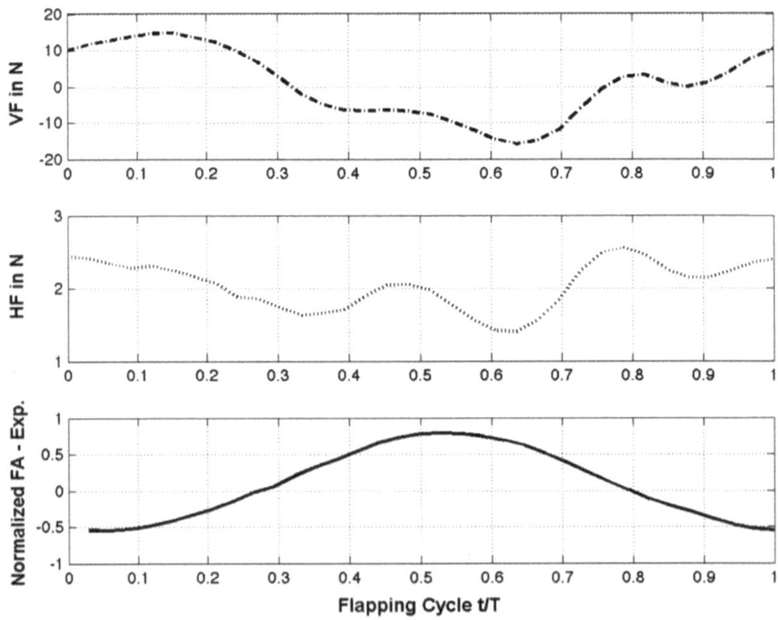

Figure 3.32 Vertical propulsive forces acting on the fuselage center of mass—obtained from experiment. *VF*, vertical force; *HF*, horizontal force; FA, flapping angle.

3.2.4 Vacuum Camber Experiment—E3

To verify inertial forces without the presence of aerodynamics, the primary test platform was tested in a vacuum. Testing was conducted at NASA Langley Research Center.

The experimental test platform was the primary test platform ML101. It was mounted on a six DoFs load cell in a vacuum chamber and integrated forces were measured. Fig. 3.33 shows the experimental setup. Reference [8] discusses further details regarding the load test.

An additional wing kinematics test was developed and conducted within the assumptions and constraint of this model [2]. Inclusion of wing kinematics data obtained in the vacuum exceeds the framework of this work.

3.2.4.1 Integrated Inertial Forces—E3

Extracted data from the test used for the verification of this model were inertial horizontal propulsive force and inertial vertical propulsive force (Fig. 3.34). Owing to the small measured magnitude of the integrated forces and the corresponding low signal-to-noise ratio mean, integrated forces were used only for verification reference. Pitching moment and verification of integrated forces over the flapping cycle were also excluded for this reason.

Experimentally measured data used and obtained for this work have been presented in this chapter, and experiments E1, E1–I, E2, E3 were discussed briefly. Data were utilized for verification of forces produced by the model,

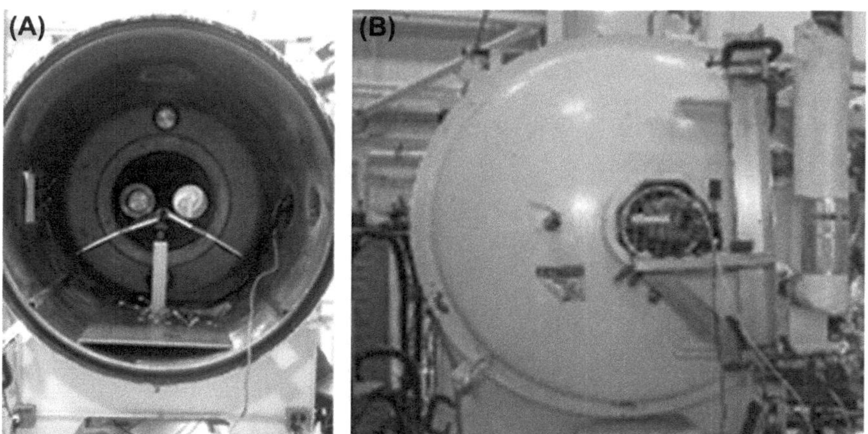

Figure 3.33 Vacuum chamber at NASA Langley Research Center. (Left) Test platform mounted on six degrees of freedom load cell, (Right) vacuum chamber.

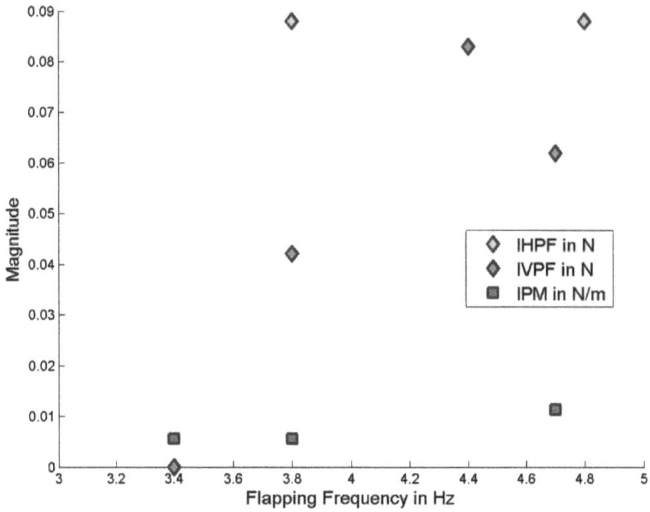

Figure 3.34 Results of vacuum chamber test ML101 test platform versus flapping frequency, (A) inertial horizontal propulsive force IHPF (B) inertial horizontal propulsive force (C) inertial pitching moment (IPM) versus magnitude.

coupled and decoupled, for verification of kinematics simulation, final model initialization, input kinematics for the aerodynamics model, and to check structural and flexible multi-body dynamics modeling assumptions.

REFERENCES

[1] W. Shyy, Y. Lian, J. Tang, D. Viieru, H. Liu, Aerodynamics of Low Reynolds Number Flyers, Cambridge University Press, New York, NY, 2008.

[2] C. Altenbuchner, J.E. Hubbard, A.A. Wissa, Free flight validation of a flexible-multi-body structural dynamics model of a bioinspired ornithopter, in: 51st AIAA Aerospace Sciences Meeting, January 2013.

[3] E. Avadikian, J. Hubbard, Ornithopter Development and Optimization Guide, Tech. Rep., 2010.

[4] R.L. Harmon, J.E. Hubbard, Aerodynamic Modeling of a Flapping Membrane Wing Using Motion Tracking Experiments (Master's Thesis), 2008.

[5] J. Grauer, J.E. Hubbard, Modeling and System Identification of an Ornithopter Flight Dynamics Model, Dissertation University of Maryland College Park, ProQuest Dissertations Publishing, 2012, 3517533.

[6] J. Grauer, J. Hubbard, Modeling of ornithopter flight dynamics for state estimation and control, in: IEEE, American Controls Conference, June 2010.

[7] A. Wissa, N. Guerreiro, J.A. Grauer, J. Hubbard, M. Frecker, C. Altenbuchner, Y. Tummala, R. Roberts, Flight testing of novel compliant spines for passive wing morphing on ornithopters, in: Structural Dynamics, and Materials and Co-located Conferences, April 2013.

[8] A. Wissa, Analytical Modeling and Experimental Evaluation of a Passively Morphing Ornithopter Wing, University of Maryland, Department of Aerospace Engineering, 2014.

CHAPTER FOUR

Flexible Multi-Body Dynamics Modeling Methodology Implementation Avian Scale Flapping Wing Flyer

Flexible multi-body dynamics is a multi-disciplinary field that encompasses several other subjects such as rigid-body dynamics, continuum mechanics, finite element method, as well as numerical, matrix, and computer methods.

Basic approaches used in simulation and computer-aided calculations for flexible multi-body dynamics and kinematics can be categorized into four different approaches: the floating frame of reference (FFR) formulation, the finite element incremental method, the large rotation vector formulations, and the absolute nodal coordinate formulation (ANCF). Approaches such as the finite segment method and the linear theory of elastodynamics (LTED) can be considered as the derivatives of the first three basic methods mentioned above [1].

The field of flexible multi-body dynamics has advanced significantly in the last two decades. However, the desired level of maturity to apply one numerical method to any flexible multi-body dynamics problem is still lacking and a deep understanding of the physics of the modeled system is required to use appropriate techniques and accomplish reliable calculations.

The field of flexible multi-body dynamics can be furthered significantly if the relationships between different formulations are established. This allows the assumptions and approximations underlying each method to be clearly defined, and research is needed in this area [1]. The potential method for all flexible multi-body dynamics problems is the ANCF is generally thought to be able to be used for complex and highly nonlinear systems. However, it is computationally very expensive due to the high number of degrees of freedom (DOFs) and therefore a challenge in itself to be applicable for complex systems.

A wide range of DOF reduction techniques exist and are a current research topic of interest. Earlier attempts used static condensation techniques, i.e., Guyan reduction, which rely on master and slave nodes [2,3]. Reduction techniques only consider stiffness properties and often fail

Modern Flexible Multi-Body Dynamics Modeling Methodology for Flapping Wing Vehicles
ISBN: 978-0-12-814136-6
http://dx.doi.org/10.1016/B978-0-12-814136-6.00004-4

to accurately predict the total mass of the system, center of mass location, moment of inertia, and natural frequencies. The work presented here implemented a modal superposition method, which accounts for full inertia coupling and relates it to a rigid multi-body assumption of the problem. In this approach, a large number of DOFs of a flexible-body are approximated as a linear combination of a smaller number of shape vectors and integrated as such in the global system.

The ornithopter at hand resulted in a linear elastic multi-body dynamic system in floating reference frame formulation and has the following properties discussed in this chapter. This development in conjunction with experimental data leads to a linear elastic multi-body system using five bodies. The floating reference frame formulation resulted in two sets of coordinates to describe the configuration of the individual deformable wing bodies. The elastic deformations are described in local wing fixed reference frames. The wing fixed reference frames are actuated to achieve large global motions, i.e., large flapping motion and oscillating body dynamics. The structural deformations are obtained by a linear combination of mode shapes. The flexible body constraints and modes are implemented using the Craig–Bampton method. Background, methods used, assumptions, and equation of motion (EOM) used for the ornithopter model are described in this section.

4.1 LINEAR ELASTIC MULTI-BODY SYSTEMS

EOMs for the ornithopter-multi-body system were solved using three formulations. First, a rigid-body formulation and an approach which is coherent with the LTED by using incremental finite element formulations that is the decoupling of the structural and rigid-body EOMs. Second, a formulation using full coupling of the EOMs through a reduced model compared to an ANCF. This is an appropriate FFR formulation. EOMs for methods are discussed and investigated in the following in terms of suitability for the flexible multi-body dynamics simulation of ornithopter. The final EOMs used for the ornithopter system is described in the following paragraphs. The approaches here were also reviewed in terms of aerodynamic coupling. Results suggest the inclusion of flexible cross-terms in the time-varying mass matrix of the EOM of the system (discussed in Section 6.4.3). Using the appropriate coordinate definition of the linear elastic multi-body system, the problem at hand is adequately modeled using floating frame of reference formulation (FFR) placement of the coordinate systems facilitates a stiffness matrix that is decoupled from the rigid-body EOMs.

The EOMs of a deformable body can be described in a general form according to

$$M^i\ddot{\xi}^i + K^i\xi^i = f_e^i + f_v^i + f_c^i \tag{4.1}$$

where M is the mass matrix, K is the stiffness matrix, i denotes the number of f_e bodies, the externally applied forces f_v the Coriolis and centrifugal forces, f_c the constraint forces.

The constraint forces f_c are in terms of the Jacobian matrix J and the Lagrange multipliers λ. The Jacobian matrix J is in terms of the constraint equations, the joint constraint, and specific motion trajectories (Eq. 4.2).

$$f_c = -J^{iT}\lambda \tag{4.2}$$

Using the floating reference formulation the generalized coordinate can be partitioned in reference coordinates r and elastic coordinates f according to the following:

$$\xi^i = \left[\xi_r^T q_f^T\right]^T \tag{4.3}$$

here ξ^i is the generalized coordinate of the flexible body i. ξ_r is the generalized coordinate of a rigid body comprised of three translational and three rotational coordinates (the Euler angles) thought to be the reference coordinates. q_f are the elastic coordinates and can be introduced using component mode methods, the finite element method, or experimental identification techniques [1]. They are the coordinates of the flexible body in the local reference frame and are described further in the sections below. The elastic coordinates q_f are obtained through the finite element method in this work. They are implemented through a component mode method and compared to experimental identified displacements.

Now the EOM of the deformable body can be expanded according to the following general form:

$$\begin{bmatrix} M_{rr} & M_{rf} \\ M_{fr} & M_{ff} \end{bmatrix}\begin{bmatrix} \ddot{\xi}_r \\ \ddot{q}_f \end{bmatrix} + \begin{bmatrix} 0 & 0 \\ 0 & K_{ff} \end{bmatrix}\begin{bmatrix} \xi_r \\ q_f \end{bmatrix} = \begin{bmatrix} f_{er} \\ f_{ef} \end{bmatrix} + \begin{bmatrix} f_{vr} \\ f_{vf} \end{bmatrix} + \begin{bmatrix} f_{cr} \\ f_{cf} \end{bmatrix} \tag{4.4}$$

Here the mass matrix is highly nonlinear. Inertia coupling between the reference motion and the elastic deformation occurs. The stiffness matrix is the same as the stiffness matrix used in structural dynamics due to the fact that the elastic coordinates are defined with respect to the local body coordinate system.

One can rewrite the EOM (Eq. 4.4) in a FFR formulation according to the following:

$$M_{rr}^{i}\ddot{\xi}_{r}^{i} + M_{rf}^{i}\ddot{q}_{f}^{i} = \left(f_{e}^{i}\right)_{r} + \left(f_{v}^{i}\right)_{r} + \left(f_{c}^{i}\right)_{r} \tag{4.5}$$

$$M_{fr}^{i}\ddot{\xi}_{r}^{i} + M_{ff}^{i}\ddot{q}_{f}^{i} + K_{ff}^{i}q_{f}^{i} = \left(f_{e}^{i}\right)_{f} + \left(f_{v}^{i}\right)_{f} + \left(f_{c}^{i}\right)_{f} \tag{4.6}$$

Assume the theory of linear elastodynamics the cross-terms between flexible and rigid-body modes vanish according to Eqs. (4.7) and (4.8).

$$M_{rr}^{i}\ddot{\xi}_{r}^{i} = \left(f_{e}^{i}\right)_{r} \tag{4.7}$$

$$M_{ff}^{i}\ddot{q}_{f}^{i} + K_{ff}^{i}q_{f}^{i} = \left(f_{e}^{i}\right)_{f} + \left(f_{v}^{i}\right)_{f} - M_{fr}^{i}\ddot{\xi}_{r}^{i} \tag{4.8}$$

It is assumed that the elastic deformation does not have a significant effect on the rigid-body displacements and the EOMs can be solved independently. It is shown in Eq. (4.7) that all terms, including the inertia tensor and forces, are assumed to be independent of the elastic deformations [1].

Aerodynamic loads are introduced in the equation set Eqs. (4.5) and (4.6) and equation set Eqs. (4.7) and (4.8) through external forces f_e. They are dependent on the generalized coordinate of the rigid body ξ_f as well as the modal coordinate q_f of the flexible body $f_e(\xi_f, q_f)$ and couple the EOMs.

If the inertial forces due to the elastic deformations are neglected according to Eqs. (4.7) and (4.8), the rigid-body motion and the EOMs for the elastic deformation are only coupled through the aerodynamics of the system, hence aeroelasticity. One can observe that the Eqs. (4.7) and (4.8) can be solved independently.

Eq. (4.7) can be solved for the reference coordinates, velocities, and accelerations, as well as the reaction forces using rigid multi-body computations only. Here, the obtained coordinates, velocities, and forces are substituted into Eq. (4.8) to determine the deformation of the bodies using standard finite element techniques.

Typically, in flight dynamics, the nonlinear inertial coupling becomes significant if one of the following statements is true [4]. The aerodynamic loads are small compared with inertial loads, the elastic frequencies and resulting rotational rates are of the same order, and/or the system results in a complex model geometry, such that transverse deflections cause changes in mass distribution [4].

In many of these systems, the inertia coupling between the rigid-body motion and the elastic deformation is significant such that the effect of this coupling cannot be neglected in the dynamic simulations [1]. In various applications, the elastic modes can also lead to instability of the rigid-body modes [1].

The EOMs for the flexible multi-body system of the ornithopter were solved using both systems of Eqs. (4.5) and (4.6) and Eqs. (4.7) and (4.8), respectively. The accuracy of the results obtained using the LTED according to Eqs. (4.7) and (4.8) is questionable, particularly, when high-speed, light-weight mechanical systems are considered such as the ornithopter system at hand [1].

EOM resolutions differ in the inclusion of the flexible cross-terms in the time-varying mass matrix and are discussed next. Three cases were implemented to solve mentioned EOMs and compared in the five-body system of the ornithopter: rigid-body motion only, no inertial coupling between rigid and flexible body EOMs, and full coupling.

Continuing the notation of the time-varying mass matrix for the case *rigid-body motion only* becomes:

$$M = \begin{bmatrix} M_{rr} & 0 \\ 0 & 0 \end{bmatrix} \tag{4.9}$$

Solving the EOMs independently and coupling only occurs through the aerodynamic forces. The flexible body's inertial properties are unaffected by deformation, that is, deformation and rigid-body motion are uncoupled [5]. The time-varying mass matrix becomes the following:

$$M = \begin{bmatrix} M_{rr} & 0 \\ 0 & M_{ff} \end{bmatrix} \tag{4.10}$$

And the final case, *full coupling of inertia terms* according to the following mass matrix:

$$M = \begin{bmatrix} M_{rr} & M_{rf} \\ M_{fr} & M_{ff} \end{bmatrix} \tag{4.11}$$

The variation of inertia terms is implemented using the Craig–Brampton method described later. The expression for the mass matrix of the generalized coordinate $M(\xi)$ simplifies to an expression in *nine* inertia invariants also further described in the EOMs (Eqs. 4.82−4.90). Here the subscripts "*t*" denote translational, *o* rotational, and *m* modal DOFs. The implemented mass matrix in terms of the generalized coordinate corresponds to the matrix terms introduced above according to the following:

For the first case, the time-varying mass matrix becomes,

$$M(\xi) = \begin{bmatrix} M_{tt} & M_{to} & 0 \\ M_{to}^T & M_{oo} & 0 \\ 0 & 0 & 0 \end{bmatrix} = \begin{bmatrix} M_{rr} & 0 \\ 0 & 0 \end{bmatrix} \tag{4.12}$$

For the second case, the implemented time-varying mass matrix corresponds to the following terms:

$$M(\xi) = \begin{bmatrix} M_{tt} & M_{to} & 0 \\ M_{to}^T & M_{oo} & 0 \\ 0 & 0 & M_{mm} \end{bmatrix} = \begin{bmatrix} M_{rr} & 0 \\ 0 & M_{ff} \end{bmatrix} \tag{4.13}$$

Finally, for the fully coupled case, the implemented time-varying mass matrix corresponds to following parts of the flexible and rigid-body cross-terms,

$$M(\xi) = \begin{bmatrix} M_{tt} & M_{to} & M_{tm} \\ M_{to}^T & M_{oo} & M_{om} \\ M_{tm}^T & M_{om}^T & M_{mm} \end{bmatrix} = \begin{bmatrix} M_{rr} & M_{rf} \\ M_{fr} & M_{ff} \end{bmatrix} \tag{4.14}$$

The FFR formulation results in a highly nonlinear mass matrix because of the inertia coupling between the reference motion and the elastic deformation [1]. Results of the comparison of the three cases suggest inclusion of flexible cross-terms in the highly nonlinear mass matrix and are presented in Section 6.4.3.

The stiffness matrix now becomes,

$$K(\xi) = \begin{bmatrix} K_{tt} & K_{to} & K_{tm} \\ K_{to}^T & K_{oo} & K_{om} \\ K_{tm}^T & K_{om}^T & K_{mm} \end{bmatrix} = \begin{bmatrix} 0 & 0 \\ 0 & K_{ff} \end{bmatrix} \tag{4.15}$$

Other than the time-varying mass matrix, the stiffness matrix is the same as the stiffness matrix used in structural dynamics. This is due to the fact that the elastic coordinates are defined with respect to the body coordinate system; the stiffness matrix is according to the following (Eq. 4.16) [1].

$$K(\xi) = \begin{bmatrix} 0 & 0 & 0 \\ 0 & 0 & 0 \\ 0 & 0 & K_{mm} \end{bmatrix} = \begin{bmatrix} 0 & 0 \\ 0 & K_{ff} \end{bmatrix} \tag{4.16}$$

4.1.1 A Floating Frame of Reference Formulation

The FFR formulation is the state-of-the-art and most widely used method in computer-aided simulations of flexible multi-body systems. It has been widely established to obtain efficient and accurate solution for low-order models. Finite element FFR formulation is used, for the most part, in the small deformation analysis of flexible bodies that undergo large reference displacements [6]. This formulation allows for filtering out systematically complex shapes associated with high frequencies that have no significant effect on the solution in the case of small deformations, therefore the formulation of a linear elastic multi-body system used in this work [6].

In the FFRs, two sets of coordinates are used to describe the configuration of the deformable bodies. One set describes the location and orientation for a selected body coordinate system ξ_r, while the other set describes the deformation of the body with respect to its coordinate system q_f.

The generalized coordinates of the linear elastic multi-body systems are used as follows:

$$\xi = \left[\xi_r^T q_f^T\right]^T$$

Model verification concluded that with the appropriate choice of coordinate systems, the FFR formulation can be used to describe the ornithopter appropriately [1,6].

4.2 THE FIVE-BODY MULTI-BODY DYNAMICS MODEL

The linear elastic multi-body system of the ornithopter and model iteration and verification resulted in a five-body multi-body system dynamics model of the ornithopter according to Fig. 4.1.

Body I, the ornithopter fuselage body is formulated by using a rigid-body formulation, so two flexible bodies connected to each other comprise each wing. Label II denotes the luff region body, and label III the thrust flap region body. Label R denotes the right wing, Label L the left wing. A flapping motion is superimposed between the Bodies I and II, its position is the flapping angle beta (β) in the following. A further flapping motion

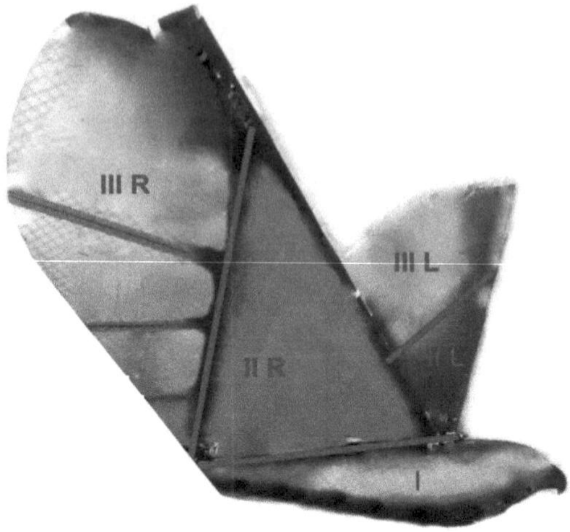

Figure 4.1 Schematic five-body dynamics system.

is superimposed between the Bodies II and III, its position is denoted the thrust flap angle zeta (ζ) in the following.

A schematic is presented to derive EOMs in this chapter. Floating reference frame multi-body dynamics formulation results into the following coordinate systems described in the following section.

4.3 RELEVANT COORDINATE SYSTEMS

Seven coordinate systems are necessary to derive the EOMs of the flexible multi-body ornithopter system and used to derive the aerodynamic model as well. The coordinate systems are defined according to Figs. 4.2 and 4.3.

The EOMs are described in the inertial fixed reference frame CI. CB is a body fixed reference frame at the fuselage center of mass. For convenience, in terms of aerodynamic model formulation and verification with experimental data, a fuselage body fixed reference frame CB0 was attached to the wing root.

Local wing fixed reference frames ZBW and ZBT were attached to the wing according to Fig. 4.3. ZBW was attached to the wing root, and ZBT to the thrust flap region. The wing fixed reference frames undergo nonlinear global motions, and the linear elastic deformations were represented in the local wing fixed reference frames.

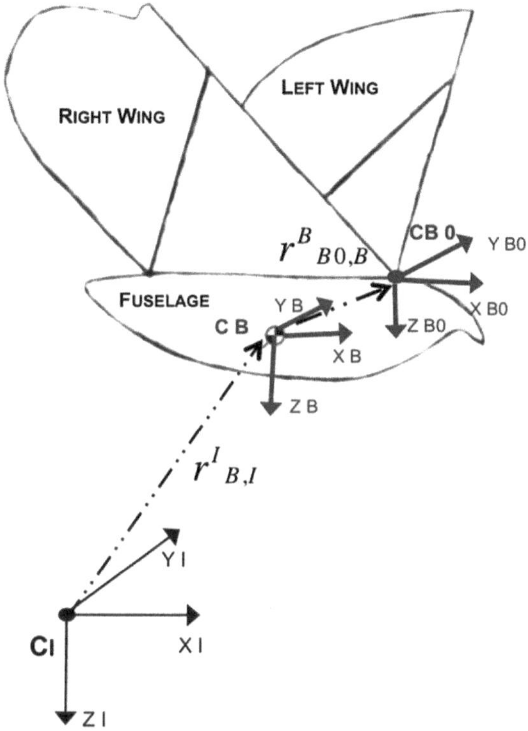

Figure 4.2 Schematic inertial and fuselage body fixed reference frames.

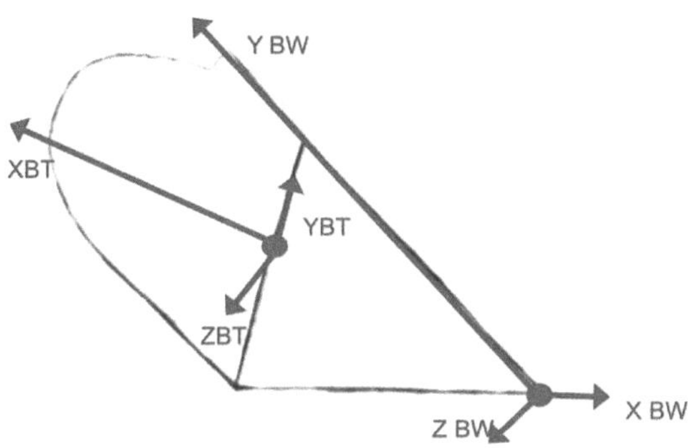

Figure 4.3 Schematic wing fixed reference frames.

4.4 AN UNDERLYING ARTICULATED RIGID-BODY MODEL

The underlying rigid motion resulting from model iteration and comparison to experimental data is described here. The model was extended to the flexible multi-body representation. In the following, the EOMs for a rigid-body structure of ornithopter are developed.

The choice of rigid-body links and connections in Fig. 4.4 is consistent with the underlying rigid-body motions of the flexible multi-body system of an ornithopter but is discussed in a broader sense.

An articulated rigid-body system is considered according to Fig. 4.4. It consists of a set of rigid bodies connected through joints in a treelike structure. Exactly one parent joint connects every rigid link [7]. The joint corresponding to the root is separate, as it does not connect to any other rigid link. The generalized coordinates are therefore the DOFs of the root link of the tree and can represent the global translation and rotation. The joint angles correspond to the admissible joint rotations for all the other joints [7].

$k = 1,\ldots,m$ and m denotes the number of rigid links. $k = 5$ for the ornithopter system at hand, here $k = 1$ is Body I, $k = 2$ is Body IIL, $k = 3$ is Body IIL, $k = 4$ is Body IIR, and $k = 5$ is Body IIIR according

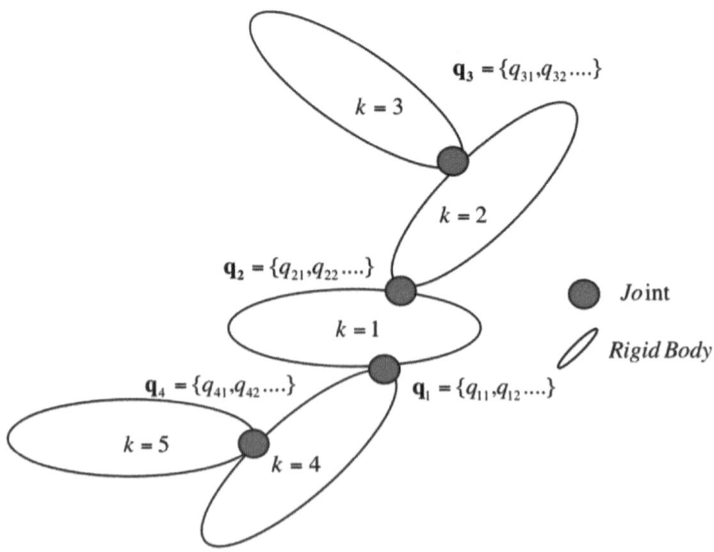

Figure 4.4 Articulated multi-body system representation of the ornithopter.

to Fig. 4.1. The state of an articulated rigid-body system as in Fig. 4.4 can be expressed in terms of $(x_k, R_k, v_k, \omega_k)$. Where x_k and R_k are the position and orientation coordinates, and v_k and ω_k are the linear and angular velocity of the center of mass of the rigid link k, respectively. They are viewed in the inertial reference frame.

The articulated rigid-body system can also be formulated in terms of generalized coordinates. One defines the generalized state as $(\mathbf{q}, \dot{\mathbf{q}})$, where $\mathbf{q} = (\mathbf{q}_1, \ldots, \mathbf{q}_k, \ldots \mathbf{q}_m)$.

Each $\mathbf{q}_k$ is the set of DOFs that connects the joint to the link k and to its parent link, i.e., for at ball joint three DOFs, a universal joint two DOFs and a hinge joint one DOFs [7].

Useful definitions for the following derivations are as follows:

$p(k)$ returns the index of the parent link k, i.e., $p(5) = 2$, link 5 has two parent links, $p(1, k)$ returns the indices of all the links in the chain from the root to the link k. For the rigid multi-body system of the ornithopter, see Fig. 4.4.

$$\begin{aligned} &p(1,1) = \{1\}, \;\; p(1,2) = \{1,2\}, \;\; p(1,3) = \{1,2,3\}, \\ &p(1,4) = \{1,4\}, \;\; p(1,5) = \{1,4,5\} \end{aligned} \qquad [7].$$

- $n(k)$ returns the number of DOFs in the joint that connects the link k to the parent link $p(k)$ [7]. For example, for the ornithopter system in Fig. 4.4 $n(1) = 6, \; n(2) = 1, \; n(3) = 1, \; n(4) = 1, \; n(5) = 1$. The total number of DOFs in the system is $n = 10$.
- R_k is the rotation matrix for the link k and depends only on the DOFs. The chain of rotational transformation from the inertial frame to the local frame of the link k is R_k^0. It follows, $R_k^0 = R_{p(k)}^0 R_k$. Since link 1 does not have a parent link and the root link represents the global translation and rotation it follows. $R_{p(1)}^0 = I_3$ [7].

4.4.1 The Kinematic Relations

The same derivation of equations as a single rigid body and for the relation between Cartesian velocities and generalized velocities is applied for the articulated rigid-body system. Here the Jacobian for each rigid link that relate its respective Cartesian velocities to the generalized velocity of the entire system is defined.

4.4.1.1 Angular Velocity

In skew-symmetric matrix form the angular velocity of k can be formulated as follows viewed in the inertial frame [7].

$$
\begin{aligned}
[\omega_k] = \dot{R}_k^0 R_k^{0T} &= \left(R_{p(k)}^0 R_k\right)\left(R_{p(k)}^0 R_k\right)^T \\
&= \left(\dot{R}_{p(k)}^0 R_k + R_{p(k)}^0 \dot{R}_k\right)\left(R_k^T R_{p(k)}^0\right)^T \\
&= \dot{R}_{p(k)}^0 R_{p(k)}^o T + R_{p(k)}^0 \left(\dot{R}_k R_k^T\right) R_{p(k)}^0 T \\
&= \left[\omega_{p(k)}\right] + R_{p(k)}^0 [\widehat{\omega}_k] R_{p(k)}^0 T
\end{aligned}
\tag{4.17}
$$

Hereby $\omega_{p(k)}$ denotes the angular velocity of the parent link in the inertial reference frame

$$
\omega_{p(k)} = \dot{R}_{p(k)}^0 R_{p(k)}^0 T \tag{4.18}
$$

and $\widehat{\omega}_k$ denotes the angular velocity of the link in the local reference frame of its parent link $p(k)$ [7].

$$
\widehat{\omega}_k = \dot{R}_k R_k^T \tag{4.19}
$$

Now the angular velocity can be reformulated in terms of the generalized state.

$$
\widehat{\omega}_k = \widehat{J}_{\omega k} \dot{q}_k \tag{4.20}
$$

where $\widehat{J}_{\omega k}$ has the dimension of is $3 \times n(k)$ and is the local Jacobian matrix that relates the joint velocity of the link k, its angular velocity in the local reference frame [parent link $p(k)$].

For skew-symmetric matrices,

$$
[R\omega] = R[\omega]R^T \tag{4.21}
$$

The angular velocity in the inertial reference frame according to Eq. (4.17) can be rewritten as follows [7].

$$
\omega_k = \omega_{p(k)} + R_{p(k)}^0 \widehat{J}_{\omega k} \dot{\mathbf{q}}_k \tag{4.22}
$$

Unrolling the recursive definition one obtains:

$$
\omega_k = J_{\omega k} \dot{\mathbf{q}}_k \tag{4.23}
$$

where the Jacobian $J_{\omega k}$ is:

$$
J_{\omega k} = \left(\widehat{J}_{\omega 1} \ldots \widehat{R}_{p(l)} \widehat{J}_{\omega l} \ldots 0 \ldots\right) \tag{4.24}
$$

The "0" corresponds to the links in the multi-body system, which is not in the chain. Also l is defined as $l \in p(1,k)$ [7].

The articulated rigid multi-body system of the ornithopter in Fig. 4.4 is as follows:

$$\omega_1 = \left(\widehat{J}_{\omega 1}\ 0\ 0\ 0\ 0\right)\dot{\mathbf{q}} \tag{4.25}$$

$$\omega_2 = \left(\widehat{J}_{\omega 1}\ R_0^1\widehat{J}_{\omega 2}\ 0\ 0\ 0\right)\dot{\mathbf{q}} \tag{4.26}$$

$$\omega_3 = \left(\widehat{J}_{\omega 1}\ R_1^0\widehat{J}_{\omega 2}\ R_2^0\widehat{J}_{\omega 3}\ 0\ 0\right)\dot{\mathbf{q}} \tag{4.27}$$

$$\omega_4 = \left(\widehat{J}_{\omega 1}\ 0\ 0\ R_1^0\widehat{J}_{\omega 4}\ 0\right)\dot{\mathbf{q}} \tag{4.28}$$

$$\omega_5 = \left(\widehat{J}_{\omega 1}\ 0\ 0\ R_1^0\widehat{J}_{\omega 4}\ R_4^0\widehat{J}_{\omega 5}\right)\dot{\mathbf{q}} \tag{4.29}$$

Angular velocities correspond to the angular velocities of the wing fixed reference frames Body I, IIR, IIL, IIIR, IIIL defined in Section 4.2, where $\widehat{J}_{\omega 1} \in IR^{3\times 3}$, $\widehat{J}_{\omega 2}$ to $\widehat{J}_{\omega 5} \in IR^{3\times 1}$.

4.4.1.2 Linear Velocity

Similar to the angular velocity, the linear velocity of the center of mass of the link k can be expressed in terms of generalized velocity [7].

$$v_k = J_{vk}\dot{\mathbf{q}}_k \tag{4.30}$$

where,

$$J_{vk} = \frac{\partial x_k}{\partial \mathbf{q}} = \frac{\partial W_k^0 c_k}{\partial \mathbf{q}} \tag{4.31}$$

The chain of homogeneous transformations from the inertial frame to the local frame of link k is denoted as W_k^0. Note that W_k^0 is different from R_k^0 in that W_k^0 includes the translational transformations. c_k is a constant vector that denotes the center of mass of link k in its local frame [7].

Now one can concentrate the Cartesian velocities into a single vector $\mathbf{V}_k^0$ and denote the relation as:

$$\mathbf{V}_k = J_{vk}\dot{\mathbf{q}}_k \tag{4.32}$$

where

$$\mathbf{V}_k = \begin{pmatrix} v_k \\ \omega_k \end{pmatrix} \text{ and } J_k = \begin{pmatrix} J_{vk} \\ J_{\omega k} \end{pmatrix} \tag{4.33}$$

4.5 LAGRANGE FORMULATION OF EQUATIONS OF MOTION

The governing EOMs for a flexible body is introduced and derived in the remainder of this section. The final EOM originates from Lagrange's equations of the following form [8].

$$\frac{d}{dt}\left(\frac{\partial L}{\partial \dot{\xi}}\right) - \frac{\partial L}{\partial \xi} + \frac{\partial C}{\partial \dot{\xi}} + \left[\frac{\partial \Psi}{\partial \xi}\right]^T \lambda = Q \tag{4.34}$$

Here T denotes the kinetic energy, V the potential energy, C is the energy dissipation function, Ψ is the algebraic constraint equation, λ is the Lagrange multipliers for the constraint, ξ is the generalized coordinates as defined in Eq. (4.47). Q is the generalized applied forces and represent the applied forces projected on ξ. Where the algebraic constraint equation is 0 according to

$$\Psi = 0 \tag{4.35}$$

Here L is the Lagrange the difference between kinetic energy and potential energy according to

$$L = T - V \tag{4.36}$$

4.5.1 Kinetic Energy Formulation

The kinetic energy in terms of the generalized coordinate ξ becomes

$$T = \frac{1}{2}\dot{\xi}M(\xi)\dot{\xi} \tag{4.37}$$

where the petitioned mass matrix is according to above definition [5].

$$M(\xi) = \begin{bmatrix} M_{tt} & M_{to} & M_{tm} \\ M_{to}^T & M_{oo} & M_{om} \\ M_{tm}^T & M_{om}^T & M_{mm} \end{bmatrix} \tag{4.38}$$

Terms in the mass matrix in terms of the inertia invariants and the position vector are further described after the derivation of the EOMs in equations for the five-body system in Eqs. (4.75)—(4.80).

4.5.2 Potential Energy Formulation

The potential energy in terms of the generalized coordinate ξ becomes,

$$V = V_g(\xi) + \frac{1}{2}\xi^T K\xi \tag{4.39}$$

Recall the generalized stiffness matrix, which becomes according to the following equation,

$$K(\xi) = \begin{bmatrix} K_{tt} & K_{to} & K_{tm} \\ K_{to}^T & K_{oo} & K_{om} \\ K_{tm}^T & K_{om}^T & K_{mm} \end{bmatrix} = \begin{bmatrix} 0 & 0 & 0 \\ 0 & 0 & 0 \\ 0 & 0 & K_{mm} \end{bmatrix} \tag{4.40}$$

It is generally a constant and only the modal coordinate contributes to the elastic energy [5]. It is not the full structural stiffness matrix of the component. $V_g(\xi)$ is the gravitational potential energy.

The governing differential EOM of a multi-body system, in terms of the generalized coordinates, can be derived in the following general form:

$$M(\xi)\ddot{\xi} + \dot{M}(\xi)\dot{\xi} - \frac{1}{2}\left[\frac{\partial M(\xi)}{\partial \xi}\dot{\xi}\right]^T \dot{\xi} + K\xi + f_g + \left[\frac{\partial \Psi}{\partial \xi}\right]^T \lambda = Q(\xi) \tag{4.41}$$

where M is the time-varying mass matrix M, K is the generalized stiffness matrix. f_g is the generalized gravitational force, Ψ are the algebraic constraint equations, λ are the Lagrange multipliers for the constraints, ξ are the generalized coordinates of the individual body.

Here the first term denotes the position and time-varying inertial forces, the second and third term denotes the Coriolis and centrifugal forces, respectively, which can be reformulated according to the following (Eq. 4.42) [3].

$$\dot{M}(\xi)\dot{\xi} - \frac{1}{2}\left[\frac{\partial M(\xi)}{\partial \xi}\dot{\xi}\right]^T \dot{\xi} = C(\xi, \dot{\xi}) \tag{4.42}$$

$K\xi$ the forth term in Eq. (4.41) is the stiffness matrix times the generalized coordinate and equals zero for a rigid-body system.

The sixth term in the EOM (Eq. 4.41) is the constraint forces and can be formulated as follows [3].

$$\left[\frac{\partial \Psi}{\partial \xi}\right]^T \lambda = N(\xi) \tag{4.43}$$

$Q(\xi)$ in Eq. (4.41) is the generalized for applied force projected to the generalized coordinate and such as motor torque and aerodynamic forces [3].

Eq. (4.41) can now be reformulated into the following general form:

$$M(\xi)\ddot{\xi} + C(\xi, \dot{\xi}) + K\xi + f_g + N(\xi) = Q(\xi) \tag{4.44}$$

The generalized coordinate vector in the EOM is comprised of the generalized coordinate of each individual body i, where i denotes the number of bodies in the multi-body system.

$$\xi = \begin{bmatrix} \xi_1 \\ \xi_2 \\ \vdots \\ \xi_i \end{bmatrix}, \tag{4.45}$$

For a rigid body, the individual generalized coordinate is comprised of three translational coordinates and three rotational coordinates that is the Euler angles.

$$\xi_i = \begin{bmatrix} x \\ y \\ z \\ \phi \\ \theta \\ \psi \end{bmatrix}_i \tag{4.46}$$

For a flexible body i the generalized coordinate vector is additionally augmented with the modal coordinates, $\mathbf{q}_i = q_{n(n=1\ldots N)}$ where N is the number of modes.

$$\xi_i = \begin{bmatrix} x \\ y \\ z \\ \phi \\ \theta \\ \psi \\ q_{n(n=1\ldots N)} \end{bmatrix} \tag{4.47}$$

To keep the generalized coordinate of a flexible body to a reasonable size, modal truncation is necessary. The component mode synthesis via Craig—Brampton method facilitates the semiphysical subspace method.

The EOMs of the multi-body system can be further evaluated looking at the position vector of the system as follows. First, the position vector of the rigid-body EOMs for an ornithopter system is presented and extended to a flexible multi-body system to distinguish both.

4.5.3 The Position Vector and Rigid-Body Equations of Motion

Now the position vector for a rigid multi-body model of ornithopter is according to Fig. 4.5.

Here *B*0 and *M*2 represent the wing attachment points to the Fuselage. Cp is the origin of a local wing fixed reference frame where marker point *P* is attached.

The instantaneous location of a marker *P* that is attached to the rigid wing body can be expressed in the inertial reference frame using the following [5]:

$$\vec{r}_{IP} = \vec{r}_{IB} + \vec{r}_{BB0} + \vec{r}_{B0P} \tag{4.48}$$

Here, $\vec{r}_{IP}$ is the position vector origin of the inertial (ground) reference frame CI to the unreformed location of marker *P* on the rigid wing body; $\vec{r}_{IB}$ is the position vector from the origin of the inertial (ground) reference frame to the origin of the local fuselage body fixed reference frame CB; $\vec{r}_{BB0}$ is the position vector from the fuselage fixed reference frame to CB to the fuselage fixed reference frame CB0; $\vec{r}_{B0P}$ is the position vector of the unreformed position of point *P* with respect to the local fuselage body reference frame of Body *B*0.

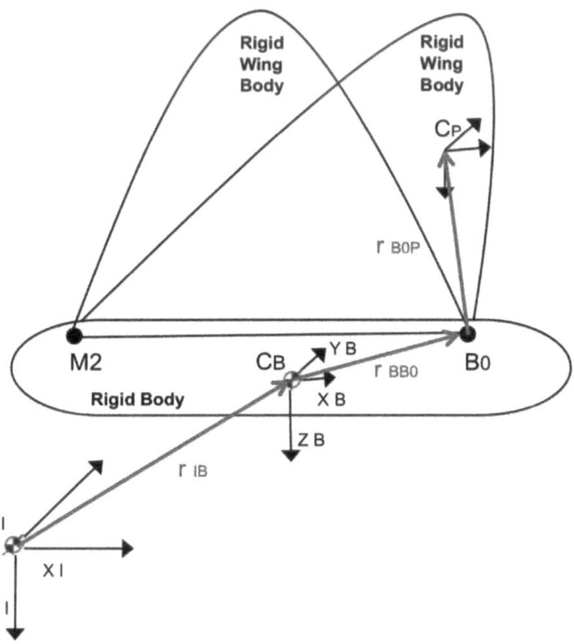

Figure 4.5 Position vector of rigid multi-body model of ornithopter.

4.5.4 The Position Vector and Flexible Body Equations of Motion

Consider a flexible wing and a point P' on the flexible wing of the ornithopter as in Fig. 4.6. P denotes a marker point on an undeformed wing body analogous to a rigid wing, P' denotes its deformed position.

The instantaneous location of the deformed point attached to a node on the flexible wing body becomes

$$\vec{r}_{IP'} = \vec{r}_{IB} + \vec{r}_{BB0} + \vec{r}_{B0P} + \vec{r}_{PP'} \tag{4.49}$$

$\vec{r}_{PP'}$ is the translational deformation vector of point P, the position vector from its undeformed to its deformed position P'.

The deformation vector is a modal superposition, for a point P/P' on a flexible body it follows

$$\vec{r}_{PP'} = \Phi_p q_n \tag{4.50}$$

The modal coordinates q_n are generalized coordinates of the flexible body according to Eq. (4.47). Φ_p is the slice from the modal matrix that corresponds to the translational DOF of node P. The dimensions of Φ_p matrix are $3 \times n$, where n is the number of modes.

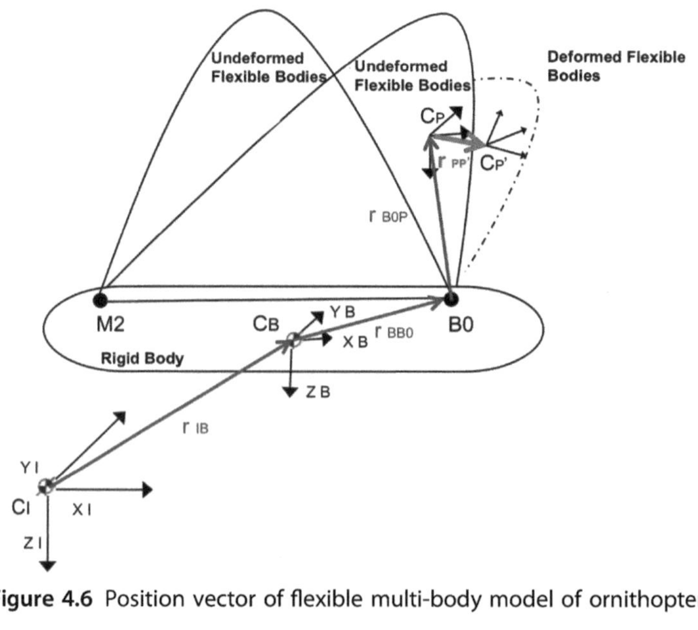

Figure 4.6 Position vector of flexible multi-body model of ornithopter.

4.5.5 The Use of Modal Superposition

The basic concept of modal superposition is that the deformation behavior of the component with a very large number of DOF such as the flexible wing can be captured with a much smaller number of modal DOFs. This reduction in the DOFs is referred to as modal truncation [9].

The discretization of the flexible wing into a finite element model (FEM) represents an infinite number of DOFs. The linear deformation of the nodes of this FEM can be approximated as a linear combination of a smaller number of shape vectors (or mode shapes), ϕ [9].

$$\mathbf{r}_{PP'} = \sum_{n=1}^{N} \phi_n q_n \tag{4.51}$$

where n is the number of mode shapes. Here Eq. (4.51) represents the scale factors or amplitudes and is the modal coordinates. Eq. (4.51) in matrix form becomes

$$\mathbf{r}_{PP'} = \mathbf{\Phi q} \tag{4.52}$$

where $\mathbf{q}$ is the vector of modal coordinates and the modes ϕ_n have been deposited in the columns of the modal matrix, $\mathbf{\Phi}$. After modal truncation, $\mathbf{\Phi}$ becomes a rectangular matrix. The modal matrix $\mathbf{\Phi}$ is the transformation for the small set of modal coordinates $\mathbf{q}$, to the larger set of physical coordinates, $\mathbf{r}_{PP'}$.

4.6 FORMULATION OF FIVE-BODY FLEXIBLE MULTI-BODY DYNAMICS MODEL

For the remainder of this section the derivation of the contributions to the final EOM of the five-body flexible multi-body system (Eq. 4.8) is discussed in the following order: (1) model generalized coordinate vector, (2) position vector, (3) velocity, (4) orientation, and (5) angular velocity.

4.6.1 The Model Generalized Coordinate Vector

The location of reference systems, body's notation, and attachment point's schematic of the model are defined (Fig. 4.7), to derive the generalized coordinate vector resulting for the five-body flexible dynamics model. Points $M1$ and T are the interface points between the bodies. The $M1$'s are located at the origin of the wing fixed reference frames BW on top

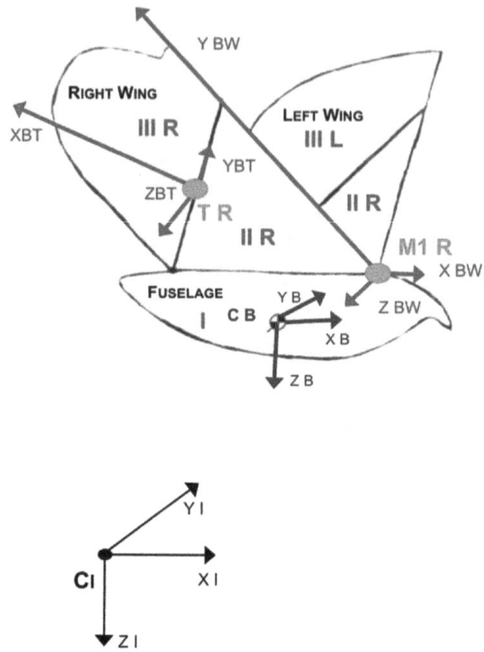

Figure 4.7 Schematic notation definition for generalized coordinates of the flexible multi-body system.

of *B*0, which is attached to the fuselage. Reference frames *BW* are attached at each wing in the luff region. *T* is located at the origin of the local body fixed thrust flap region reference frames *BT*. Since wings can be treated separately, the right and the left wings have their own attachment points denoted as *R* and *L*, respectively. Location of points *TR* and *M*1*R* is shown in Fig. 4.7.

The generalized coordinate vector of the five-body flexible multi-body dynamics systems is augmented as follows and used in the EOMs for the flexible multi-body system of ornithopter according to Eq. (4.41).

$$\xi = \begin{Bmatrix} \xi_{\mathrm{I}} \\ \xi_{\mathrm{II}R} \\ \xi_{\mathrm{II}L} \\ \xi_{\mathrm{III}R} \\ \xi_{\mathrm{III}L} \end{Bmatrix} \tag{4.53}$$

where the generalized coordinates of the bodies are according to the following:

The rigid fuselage Body I,

$$\xi_{\mathrm{I}} = \begin{Bmatrix} x_{IB} \\ y_{IB} \\ z_{IB} \\ \psi_{IB} \\ \theta_{IB} \\ \phi_{IB} \end{Bmatrix} \tag{4.54}$$

The flexible luff region wing Bodies II (R/L)

$$\xi_{\mathrm{II}(R/L)} = \begin{Bmatrix} x_{IM1} \\ y_{IM1} \\ z_{IM1} \\ \psi_{IM1} \\ \theta_{IM1} \\ \phi_{IM1} \\ q_{\mathrm{II}} \end{Bmatrix}, \tag{4.55}$$

with $q_{\mathrm{II}} = q_n(n = 1\ldots N)$.

The flexible luff region wing Bodies III (R/L)

$$\xi_{\mathrm{III}(R/L)} = \begin{Bmatrix} x_{IT} \\ y_{IT} \\ z_{IT} \\ \psi_{IT} \\ \theta_{IT} \\ \phi_{IT} \\ q_{\mathrm{III}} \end{Bmatrix} \tag{4.56}$$

with $q_{\mathrm{III}} = q_n(n = 1\ldots N)$.

4.6.2 Position Vector in the Model

For the five-body dynamics system the instantaneous location of the deformed point attached to a node on the flexible wing Body II and III is presented in the following.

The instantaneous velocity of flexible Bodies II (*R*/*L*) can be described in (Eq. 4.49) Fig. 4.6. An additional vector and therefore coordinate transformation is necessary to compute flexible points on the thrust flap region. To reach the local reference frame in which the flexible point is located on the thrust flap region an additional vector is used and is shown in Fig. 4.8.

The instantaneous velocity of flexible Bodies III (*R*/*L*) becomes

$$\vec{r}_{IP'} = \vec{r}_{IB} + \vec{r}_{BB0} + \vec{r}_{B0TR} + \vec{r}_{TRP} + \vec{r}_{PP'} \tag{4.57}$$

where the additional vector, $\vec{r}_{B0TR}$ is the position vector from the fuselage fixed reference frame to B0 to the wing fixed reference frame *TR* and, $\vec{r}_{TRP}$ is the position vector of the undeformed position of point *P* with respect to the wing fixed reference frame of the wing Body II.

The deformed position or nodal point m on the FEM of the wing luff region [Bodies II (*L*/*R*)] can now be expressed in the inertial coordinate frame according to

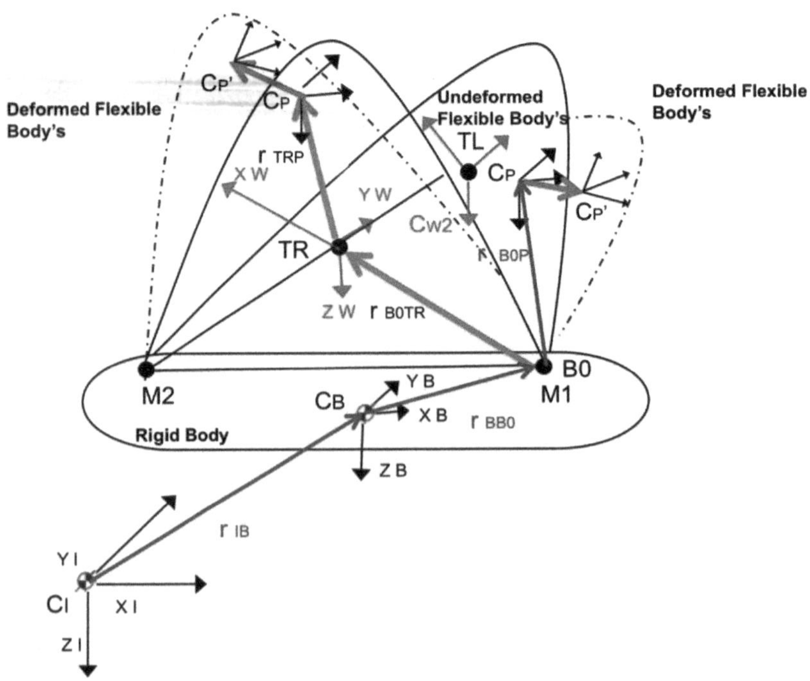

Figure 4.8 Position vector of flexible multi-body model of ornithopter, luff region [blue (darker gray in print versions) and red (light gray in print versions)], thrust flap region [orange (dark gray in print versions) and red (lighter gray in print versions)].

$$r_{IP'_m} = r_{IB0} + {}^{I}A^{B0}\left(\vec{r}_{M_1P_m} + \vec{r}_{P_mP'_m}\right) \tag{4.58}$$

where ${}^{I}A^{B0}$ is the transformation matrix from the local body reference frame of $B0$ (fuselage body fixed) to the inertial reference frame. This matrix can also be described as the direction cosines of the local body reference frame with respect to the ground. In applied Adams solver, the orientation is captured using a body fixed 3-1-3 sets of Euler angles ψ, θ, ϕ. The Euler angles are generalized coordinates of the flexible body [9].

The deformed position, or nodal point m on the FEM of the thrust flap region [Bodies III (L/R)] expressed in the inertial coordinate frame is according to Eq. (4.59).

$$\vec{r}_{IP'_m} = \vec{r}_{IB0} + {}^{I}A^{B0}(\vec{r}_{B0Pm}) \tag{4.59}$$

with,

$$\vec{r}_{B0Pm} = \vec{r}_{B0TR} + {}^{B0}A^{T}\left(\vec{r}_{TP_m} + \vec{r}_{P_mP'_m}\right) \tag{4.60}$$

where ${}^{B0}A^{T}$ is the transformation matrix from the local wing body reference frame of ZBW (luff region fixed reference frame) to the fuselage body fixed reference frame $B0$.

4.6.3 Velocity in the Model

The instantaneous translational velocity of point P' is computed by differentiating Eq. (4.58) for the luff region [Bodies II (L/R)] and Eq. (4.64) for the thrust flap region [Bodies III (L/R)], respectively. This is for the purpose for obtaining the kinetic energy of the point [9].

The instantaneous velocity for the luff region now becomes,

$$\vec{v}_{IP'_m} = \dot{\vec{r}}_{IB0} + {}^{I}\dot{A}^{B0}\left(\vec{r}_{M_1P_m} + \vec{r}_{P_mP'_m}\right) + {}^{I}A^{B0}\dot{\vec{r}}_{P_mP'_m} \tag{4.61}$$

The instantaneous velocity for the thrust flap region becomes,

$$\vec{v}_{B0P'_m} = \dot{\vec{r}}_{B0TR} + {}^{B0}\dot{A}^{T}\left(\vec{r}_{TP_m} + \vec{r}_{P_mP'_m}\right) + {}^{B0}A^{T}\dot{\vec{r}}_{P_mP'_m} \tag{4.62}$$

4.6.4 Model Orientation

To derive the transformation matrix, one considers a point L_{II} on the flexible body II (luff region of the wing). The orientation of point L relative to the inertial reference frame is represented by the transformation matrix,

$${}^{I}A^{LII} = {}^{I}A^{BB}A^{M1M1}A^{PP}A^{LII} \tag{4.63}$$

One considers a point L_{III} on the flexible Body III (thrust flap region of the wing) The orientation of point L relative to the inertial reference frame can be written by the transformation matrix,

$$^{I}A^{\mathrm{LIII}} = {}^{I}A^{BB}A^{M1M1}A^{TT}A^{PP}A^{\mathrm{LIII}} \tag{4.64}$$

In Eq. (4.60) the product of the first two transformation matrices,

$$^{I}A^{M1} = {}^{I}A^{BB}A^{M1} \tag{4.65}$$

Eq. (4.65) is the transformation matrix from the local wing fixed reference frame in which the elastic deformation is described, to the inertial reference frame of the system. In the flexible Body III (Eq. 4.63) the additional transformation matrix is added due to the additional global motion of the thrust flap region, and the transformation matrix to the local wing fixed reference frame becomes Eq. (4.66) here [9].

$$^{I}A^{T} = {}^{I}A^{BB}A^{M1M1}A^{T} \tag{4.66}$$

$^{I}A^{M1}$ is the transformation matrix from the local body reference frame of BW to the inertial reference frame CI, and $^{I}A^{T}$ is the transformation matrix from the local body reference frame of TW to the inertial reference frame CI.

The transformation matrices $^{M1}A^{P}$ and $^{T}A^{P}$ are due to the orientation change due to the deformation of the node P. If one examines $^{M1}A^{P}$ further that is analogous to $^{T}A^{P}$ one can expand the matrix according to the following. The direction cosines for a vector of small angles ε are [5,9],

$$^{M1}A^{P} = \begin{bmatrix} 1 & -\varepsilon_{pz} & -\varepsilon_{py} \\ \varepsilon_{pz} & 1 & -\varepsilon_{px} \\ \varepsilon_{py} & \varepsilon_{px} & 1 \end{bmatrix} = \mathbf{I} + \widetilde{\varepsilon}_{p} \tag{4.67}$$

Here the tilde denotes the skew operator, and $\mathbf{I}$ is the 3×3 identity matrix.

The orientation of the point on a flexible body is instantly evaluated, as the body deforms to satisfy angular constraints [5]. As the wing deforms, the marker rotates through small angles relative to its reference frames BW (luff region), BT (thrust flap region). Similar to translational deformations, these angles are obtained through modal superposition according to

$$\varepsilon_{P} = \mathbf{\Phi}_{p}^{*}\mathbf{q} \tag{4.68}$$

Here $\mathbf{\Phi}_{p}^{*}$ is the slice from the modal matrix that corresponds to the rotational DOFs of the node P and is of $3 \times N$ dimension, where N is the number of modes [9].

4.6.5 Angular Velocity

Now the angular velocity of a flexible body marker L is the sum of angular velocity of the body and the angular velocity due to deformation.

For flexible Bodies II the angular velocities become and are expressed in the wing body fixed reference frames.

$$\begin{aligned} {}^{I}\omega_{BW}^{LII} &= {}^{I}\omega_{BW}^{P} = {}^{I}\omega_{BW}^{B} + {}^{B}\omega_{BW}^{B0} + {}^{B0}\omega_{BW}^{P} \\ &= {}^{I}\omega_{BW}^{B} + {}^{I}\omega_{BW}^{B0} + \Phi * q \end{aligned} \tag{4.69}$$

For flexible Bodies III the angular velocities become and are expressed in the wing body fixed reference frames in the thrust flap region and become according to the following.

$$\begin{aligned} {}^{I}\omega_{BT}^{LIII} &= {}^{I}\omega_{BT}^{P} = {}^{I}\omega_{BT}^{B} + {}^{B}\omega_{BT}^{B0} + {}^{B}\omega_{BT}^{BT} + {}^{B0}\omega_{BT}^{P} \\ &= {}^{I}\omega_{BT}^{B} + {}^{B}\omega_{BT}^{B0} + {}^{B}\omega_{BT}^{BT} + \Phi * q \end{aligned} \tag{4.70}$$

Eqs. (4.69) and (4.70) are the underlying rigid-body motion rotations plus the rotation vector due to the elastic deflection. The angular velocities of the bodies relate to the time derivative of the orientation states according to the following.

For Bodies II (L/R) it follows,

$$ {}^{B}\omega_{BW}^{B0} = \mathbf{B}_{BW} * \dot{\psi}_{\text{II}(L/R)} \tag{4.71}$$

with $\psi_{\text{II}(L/R)}$ is a part of the generalized coordinate,

$$\psi_{\text{II}(L/R)} = \begin{Bmatrix} \psi_{IM1} \\ \theta_{IM1} \\ \varphi_{IM1} \end{Bmatrix}_{\text{II}(L/R)} \tag{4.72}$$

For Bodies III (L/R) it follows,

$$ {}^{B}\omega_{BT}^{BT} = \mathbf{B}_{BT} * \dot{\psi}_{\text{III}(L/R)} \tag{4.73}$$

where $\psi_{\text{III}(L/R)}$ is a part of the generalized coordinate,

$$\psi_{\text{III}(L/R)} = \begin{Bmatrix} \psi_{IT} \\ \theta_{IT} \\ \varphi_{IT} \end{Bmatrix}_{\text{III}(L/R)} \tag{4.74}$$

Here $\mathbf{B}_{BW}$ and $\mathbf{B}_{BT}$ are the respective transformation matrices.

4.6.6 Inertia Invariants in Mass Matrix

The expression for the mass matrix results in an expression of nine inertia invariants, which can now be defined using the position vector. Recall, the time-varying mass matrix,

$$M(\xi) = \begin{bmatrix} M_{tt} & M_{to} & M_{tm} \\ M_{to}^T & M_{oo} & M_{om} \\ M_{tm}^T & M_{om}^T & M_{mm} \end{bmatrix} \tag{4.75}$$

where the terms in the mass matrix are dependent on the inertia invariants in the following Eqs. (4.76)–(4.81),

$$M_{tt} = I^1 \mathbf{I} \tag{4.76}$$

$$M_{to} = -A\left[I^2 + I_j^3 q_j\right]B \tag{4.77}$$

$$M_{tm} = AI^3 \tag{4.78}$$

$$M_{oo} = B^T\left[I^7 - \left[I_j^8 + I_j^8\right]q_j - I_{ij}^9 q_i q_j\right]B \tag{4.79}$$

$$M_{om} = B^T\left[I^4 - I_j^5 q_j\right] \tag{4.80}$$

$$M_{mm} = I^6 \tag{4.81}$$

It can be observed that the mass matrix is explicit dependent on the modal coordinates. A and B represent the transformation matrices of the system [5].

The inertia invariants are computed from the nodes N of the FEM. Inertia variants are based on the node's mass m_{pp}, its undeformed location v_p, and its participation in the component modes $\mathbf{\Phi}_p$. The discrete form of the inertia invariants is provided in equations below [5].

$$I^1 = N\sum_{p=1}^{N} m_p \tag{4.82}$$

$$I^2 = \sum_{p=1}^{N} m_p v_{p(\mathrm{II/III})} \tag{4.83}$$

$$I_j^3 = \sum_{p=1}^{N} m_p \mathbf{\Phi}_p \tag{4.84}$$

$$I_j^4 = \sum_{p=1}^{N} m_p \widetilde{v}_{p(\mathrm{II/III})} \Phi_p + I_P \Phi_p^* \tag{4.85}$$

$$I_j^5 = \sum_{p=1}^{N} m_p \widetilde{\phi}_{pj} \Phi_p + I_P \Phi_p^* \tag{4.86}$$

$$I^6 = \sum_{p=1}^{N} m_p \Phi_p^T \Phi_p + \Phi_p^{*T} I_P \Phi_p^* \tag{4.87}$$

$$I^7 = \sum_{p=1}^{N} m_p \widetilde{v}_{p(\mathrm{II/III})}^T \widetilde{v}_{p(\mathrm{II/III})} + I_p \tag{4.88}$$

$$I_j^8 = \sum_{p=1}^{N} m_p \widetilde{v}_{p(\mathrm{II/III})} \phi_{pj} \tag{4.89}$$

$$I^9 = \sum_{p=1}^{N} m_p^T \widetilde{\phi}_{pj} \widetilde{\phi}_{pk} \tag{4.90}$$

With v_p is the position vector from the local body reference frame of B to the point P, expressed in the local body coordinate system, therefore it follows for Body II and Body III it follows, [2,5].

For Body II (L/R) it follows,

$$v_{P\mathrm{II}} = \vec{r}_{BB0} + \vec{r}_{B0P} \tag{4.91}$$

And Body III (L/R):

$$v_{P\mathrm{III}} = \vec{r}_{IP'} = \vec{r}_{BB0} + \vec{r}_{B0TR} + \vec{r}_{TRP} \tag{4.92}$$

4.6.7 The Craig–Brampton Method

To facilitate the modal superposition in the flexible multi-body system, a component mode synthesis method is used, which is the Craig–Brampton method.

Now the goal is to select the mode shape so the largest possible amount of deformation can be captured with a minimum number of modal coordinates. This ensures the modal basis is optimized.

It was previously assumed that eigenvectors would provide a useful modal basis [9]. However, this would require an excessive number of modes to capture the effects of attachments on the flexible body [9]. This could cause the generalized coordinate vector of a flexible body to become unnecessarily long. To prevent unwanted constraints in the system, eigenvectors of the unconstrained system were used. However, these eigenvectors provided an inadequate basis in the system-level modeling such as the ornithopter problem at hand.

To solve this problem, a component mode synthesis technique was used. The Craig—Brampton method was adopted. This method allowed a subset of DOF not to be subjected to modal superposition to be selected (usually referred to as boundary DOF, attachment DOF, or interface DOF). In the context interface, DOF will be used.

Interface DOF are preserved exactly in the Craig—Brampton modal basis. There is no loss in resolution to the interface DOFs, when higher-order modes are truncated.

The systems DOF are partitioned into two different sets of DOFs at the interface: boundary DOF, u_B, and interior DOF u_I. Two sets of modes shapes, the constraint modes and fixed-boundary normal modes, are defined below.

Constraint modes are static shapes obtain by giving each boundary DOF a unit displacement while holding all other boundary DOF fixed. The basis of constraint modes completely spans all possible motions of the boundary DOFs, with a one-to-one correspondence between the modal coordinates of the constraint modes and the displacement in the corresponding boundary DOF, $q_C = u_B$.

Fixed-boundary normal modes are obtained by fixing the boundary DOF when computing the Eigen solution. There are unlimited fixed-boundary normal modes. These modes define the modal expansion for the interior DOF. The quality of this modal expansion is proportional to the number of modes used for the flexible multi-body.

Now the relationship between the physical DOF and the Craig—Brampton modes and their modal coordinates can be formulated as follows [9].

$$u = \begin{Bmatrix} u_B \\ u_I \end{Bmatrix} = \begin{bmatrix} I & 0 \\ \Phi_{IC} & \Phi_{IN} \end{bmatrix} = \begin{Bmatrix} q_C \\ q_N \end{Bmatrix} \tag{4.93}$$

where deformation matrix, u, is composed of translational deformations $\mathbf{r}_{pp}$ and rotational deformations epsilon ε_P, which is described above and is according to the following:

$$u = \{ \mathbf{r}_{pp} \quad \varepsilon_P \} \tag{4.94}$$

In Eq. (4.93), u_B are the boundary DOFs, u_I are the interior DOFs, I and 0 are the identity and 0 matrices, respectively. Φ_{IC} are physical displacements of the interior DOF in the constraint modes, Φ_{IN} are physical displacements of the interior DOF in the normal modes, q_C are the modal coordinates of the constraint modes, q_N are the modal coordinates of the flexible boundary modes.

The generalized stiffness and mass matrices corresponding to the Craig–Brampton modal basis are obtained via a modal transformation. The stiffness transformation is

$$\widehat{K} = \Phi^T K \Phi = \begin{bmatrix} I & 0 \\ \Phi_{IC} & \Phi_{IN} \end{bmatrix}^T \begin{bmatrix} K_{BB} & K_{BI} \\ K_{IB} & K_{II} \end{bmatrix} \begin{bmatrix} I & 0 \\ \Phi_{IC} & \Phi_{IN} \end{bmatrix} = \begin{bmatrix} \widehat{K}_{CC} & 0 \\ 0 & \widehat{K}_{NN} \end{bmatrix} \tag{4.95}$$

and the mass transformation is according to the following,

$$\widehat{M} = \Phi^T M \Phi = \begin{bmatrix} I & 0 \\ \Phi_{IC} & \Phi_{IN} \end{bmatrix}^T \begin{bmatrix} M_{BB} & M_{BI} \\ M_{IB} & M_{II} \end{bmatrix} \begin{bmatrix} I & 0 \\ \Phi_{IC} & \Phi_{IN} \end{bmatrix} = \begin{bmatrix} \widehat{M}_{CC} & \widehat{M}_{NC} \\ \widehat{M}_{NC} & \widehat{M}_{NN} \end{bmatrix} \tag{4.96}$$

Here the subscripts B, N, C, and I denote internal DOF, boundary DOF, normal mode, and constraint mode, respectively. $\widehat{M}$, $\widehat{K}$ denote the generalized mass and stiffness matrix, respectively. The matrices $\widehat{M}_{NN}$, $\widehat{K}_{NN}$ are diagonal matrices associated with eigenvectors. $\widehat{K}$ is the block diagonal. There is no stiffness coupling between the constraint modes and fixed-boundary normal modes. $\widehat{M}$ is not block diagonal, because there is inertia coupling between the constraint modes and the fixed-boundary normal modes.

4.6.8 A Modal Synthesis Method and Mode Shape Orthonormalization

The Craig–Brampton method tailors the modal basis to capture the desired attachment effects and the desired level of dynamic content. However, further steps make it suitable for direct use in the dynamic system calculation.

This is because in the Craig–Brampton method the constraint modes are six rigid bodies DOF, which must be eliminated before the multi-body analysis because its own large motion rigid-body DOF is provided [9]. Also, the Craig–Brampton constraint modes cannot be eliminated because doing so would be comparable to applying a constraint on the system [9]. Furthermore, with the method the constraint modes are a result of static condensation [9]. Consequently, these modes do not translate the dynamic frequency content that they must contribute to a flexible body. Therefore, successful simulation of a nonlinear system with unknown frequency content would be unlikely [9].

These issues in the raw formulation of the Craig–Brampton method can be resolved by orthogonalizing the Craig–Bampton modes. The appropriate and verified application in the system would result in a stiffness matrix according to Eq. (4.16).

The Craig–Bampton modes are not an orthogonal set of modes, as evidenced by the fact that their generalized mass and stiffness matrix $\widehat{M}$, $\widehat{K}$ in Eqs. (4.95) and (4.96) are not diagonal [9].

By solving the eigenvalue problem,

$$\widehat{K}q = \lambda\widehat{M}q \tag{4.97}$$

eigenvectors are obtained that are arranged in a transformation matrix N, which transforms the Craig–Brampton modal basis to an equivalent, orthogonal basis with modal coordinates.

$$Nq^{*} = q \tag{4.98}$$

The effect of the superposition formula is:

$$\mathbf{u} = \sum_{n=1}^{N} \varphi_n q_n = \sum_{n=1}^{N} \varphi_n N q* = \sum_{n=1}^{N} \varphi_n^{*} q* \tag{4.99}$$

Here $q*$ are the orthogonalized Craig–Brampton modes.

The Craig–Bampton orthonormalization yields the modes of the unconstrained system, six of which are rigid-body modes, which can be disabled after the operation. Following the second Eigen solution, all modes have an associated natural frequency. Problems arising from modes contributing high-frequency content can now be anticipated [9]. Although after the orthonormalization, the removal of any mode disallows the body from adopting that particular shape. The removal of a high-frequency mode prevents the associated boundary node from moving relative to its neighbors. If the constraint mode with unknown frequency contribution would be removed, this would only prevent the boundary edge from reaching its degree of waviness [9].

The orthogonalized Craig–Brampton modes are a representation of the system and as such have a natural frequency associated with them [9]. Note that they are not the eigenvectors of the original system and cannot be exactly described physically but have the following characteristics.

The orthogonalized Craig–Bampton modes and fixed-boundary normal modes are replaced with an approximation of the eigenvectors of the unconstraint body. This is an approximation because it is based only on the Craig–Bampton modes. Out of these modes, six are usually the rigid-body modes. The constraint modes are replaced with the boundary eigenvector [9].

4.7 STRUCTURAL DYNAMICS MODEL OF THE WINGS

The FEM of each wing structure used in the five-body vehicle dynamics model (Fig. 4.9) was modeled using structures workspace in the SimXpert GUI. For the structural solution the geometry of ornithopter was imported and a finite element mesh created. The structural quadratic grid was refined until modal solutions were converged. For the wing structure, 1796 plate elements were used and the bending stiffness was adjusted to fit the physical system. The leading edge spar was rigidly attached to a shoulder joint and was pinned to the fuselage at CB0 and underwent a flapping motion and spatially distributed aerodynamic loads as described in Chapter 5 were applied. The aerodynamic loads were introduced as time-variant forcing functions on the nodes of corresponding wing structure region. Isotopic material was chosen for the membrane structure according to

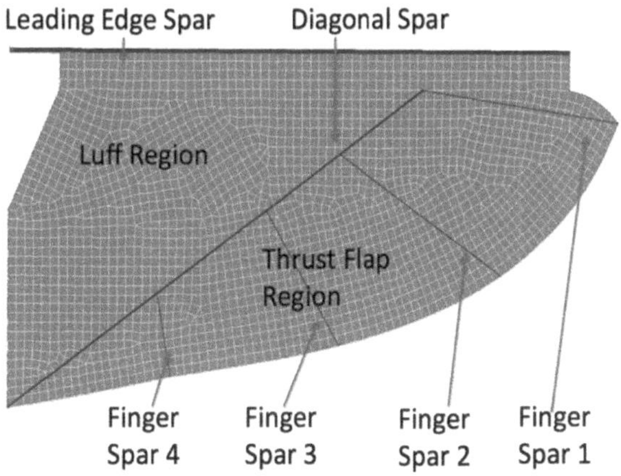

Figure 4.9 Schematic finite element model wing structure carbon fiber spars and wing.

Table 4.1 Material properties and dimensions of ornithopter structures used for the model

Element	Thickness diameter (mm)	Material	Elastic modulus (GPa)	Structural element type	Poisson ratio	Density (kg/m^3)
Wing structure	0.06	Polyester ripstop	100	Quadratic	0.31	500
Leading edge spar	3.6	Carbon fiber	230	Quadratic	0.74	1750
Diagonal spar	1.8	Carbon fiber	175	Quadratic		1750
Finger spar 1–4	1	Carbon fiber	175	Quadratic		1750

properties in Table 4.1. The CTETRA8/CTRIA6 and the CBEAM elements in SimXpert structures workspace were applied to the structure made of carbon fiber and ripstop nylon fabric skin, respectively. The beam element used for the carbon fiber spars has bending, torsion, tension, and compression properties and a one-dimensional nonlinear element with eight nodes. An elastic shell element CQUAD8 was used for the membrane structure. This element is capable of carrying in plane force, bending force, and transverse shear force. An isotropic material was used for the carbon fiber structural elements, and MAT1 and MATT1 solvers are used for the elements. Structural wing model is submitted to the solver using MD Nastran (FEM), and simulation is integrated during the multi-body dynamics simulation. In structures workspace of the SimXpert GUI modal solver, SOL103 was used to obtain as solution for the wing structure. SOL103 performs a modal analysis and was used to compute the natural frequencies and the associated mode shapes of a structure [10].

The finite element analysis model schematic of the wing and its components is shown in the following.

The wing regions Body II (*L*/*R*) and wing Body III (*L*/*R*) are divided by the diagonal spar. Each component such as the thrust flap region, the luff region, and the leading edge spar, the diagonal spar, and the finger spar 1–4 is an individual flexible element. For example, the mesh and mode shapes for the thrust flap region are shown in Figs. 4.10 and 4.11.

Finger spars 1–4 are rigidly attached to the thrust flap region, and the leading edge spar and diagonal spar are rigidly attached to the luff region.

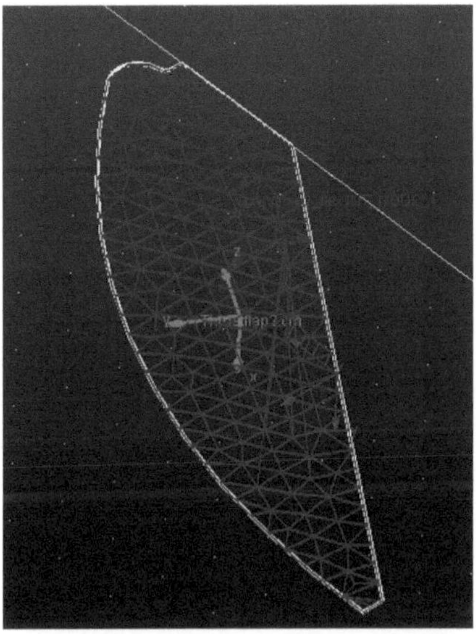

Figure 4.10 Thrust flap region mesh nodes = 352, elements = 860.

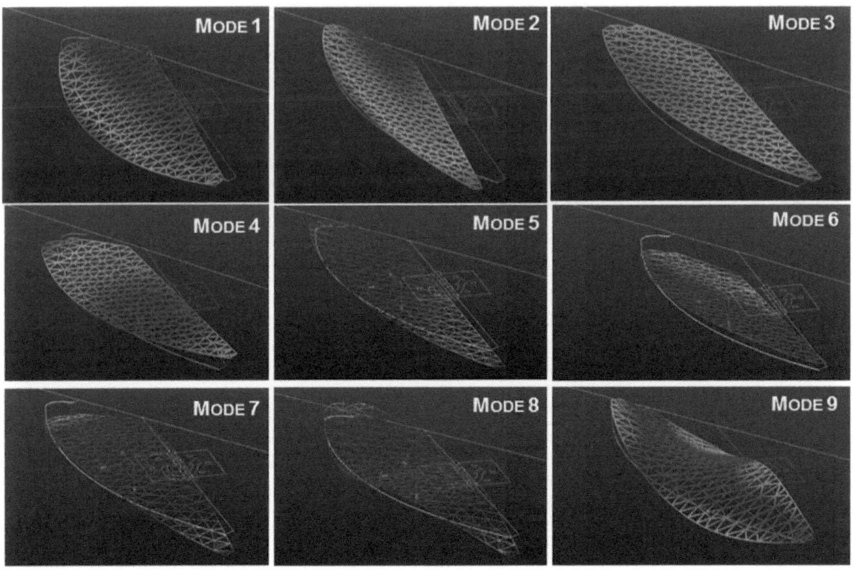

Figure 4.11 Thrust flap region mode shapes contour plot.

This results in two flexible wing bodies for each wing in the five-body flexible multi-body dynamics model. Fixed connections of flexile wing components are highlighted by lock image in five-body flexible dynamics model image Fig. 4.12.

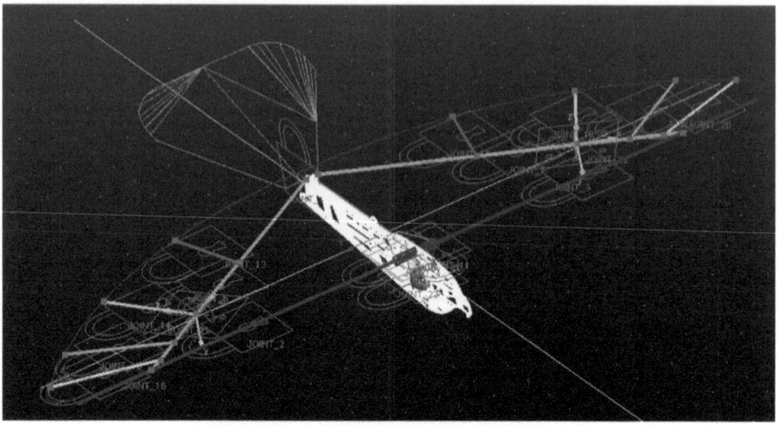

Figure 4.12 Model image flexible wing component connections.

4.7.1 Model Implementation—Mode Neutral File

The flexible structural dynamics parts are integrated in the flexible multi-body dynamics model via a modal neutral file (MNF) file. Integration was accomplished through interaction between the structures and motion workspaces in the SimXpert (GUI). A modal analysis (SOL103) was performed in the structures workspace, which generated a Nastran results file and a MNF. In the simulation, displacement and stress data are available for the body in the flexible multi-body dynamics simulation [2,5].

REFERENCES

[1] A.A. Shabana, Flexible multibody dynamics: review of past and recent developments, Multibody System Dynamics 1 (1997) 189–222.

[2] MacNeal-Schwendler Corporation (MSC), Motion Workspace Guide Introduction – Simxpert, MSC, Newport Beach, CA, 2011.

[3] MacNeil Schwendler Cooperation (MSC), Adams Flex, MSC, 2011.

[4] M.R. Waszak, C.S. Buttrill, D.K. Schmidt, NASA Technical Memoradum, Modeling and Model Simplification of Aeroelastic Vehicles: An Overview, vol. 107691, September 1992.

[5] MSC Software, Adams Flex – Reported in Adams Docs DOC10403, MSC, 2013.

[6] K. Panagoitis, Model Order Reduction in Structural Mechanics, Coupeling the Rigid and Elastic Multi Body Dynamics, Professur für Fahrzeugmodellierung und –Simulation, Institut fuer Bahnfarzeuge und Bahntechnik, Technische Universitaet Dresden, 2009.

[7] C.K. Liu, S. Jain, A Quick Tutorial on Multibody Dynamics, School of Interactive Computing Georgia Institute of Technology, 2010.

[8] J.S. Lee, J.K. Kim, DK. Kim, J.H. Han, Longitudinal flight dynamics of bioinspired ornithopter considering fluid-structure interaction, Journal of Cuidance, Control, and Dynamics 34 (3) (May–June 2011) 667.

[9] MacNeil Schwendler Cooperation (MSC), Adams. Appendix – Theoretical Background, 2008.
[10] C. Altenbuchner, J.E. Hubbard, A. Wissa, Free flight validation of a flexible-multibody structural dynamics model of a bioinspired ornithopter, in: 51st AIAA Aerospace Sciences Meeting, January 2013.

CHAPTER FIVE

Aerodynamics Modeling for Flexible Multi-Body Dynamics Modeling Methodology Implementation Avian Scale Flapping Wing Flyer

MATLAB code was written for the aerodynamic load generation in the five-body flexible multi-body dynamics model (FMBDM). The final aerodynamic model is implemented in a twofold code where a first MATLAB code was generated to extract time-variant aerodynamic loads from experimental wing kinematics. The Aero Load Experimental Free Flight (AEROLEF) code is described in the following section. After its iteration and validation, a second MATLAB code was generated to be coupled with the FMBD to simulate a different ornithopter configuration. Experimental data (E-2) serves for initialization, and kinematics from the FMBDM serve as exchange variable. The latter code, Aero Load Initialized Experimental Coupled (AEROIEC) and its formulation are also described.

5.1 THE AERODYNAMIC MODEL VERSIONS FORMULATED

Three aerodynamic models A, B, and C developed in the course of this study resulted in AEROLEF and AEROIEC. These aerodynamic models were used as time variant externally applied wing loads and differ in aerodynamic model assumptions (e.g., blade element refinement and experimental data source). Differences between models A, B, and C are specified in Table 5.1. Model C is the updated aerodynamic model with improvements to blade element refinement and aerodynamic modeling assumptions. The codes AEROLEF and AEROIEC used model C assumptions and are discussed in detail below.

Aero-models A, B, and C generate external aerodynamic forces applied to corresponding nodes in the wing model in the special aperture of a blade element, as seen in Figs. 5.1, 5.2A and B. Experimental data (E1-I) were

Modern Flexible Multi-Body Dynamics Modeling Methodology for Flapping Wing Vehicles
ISBN: 978-0-12-814136-6
http://dx.doi.org/10.1016/B978-0-12-814136-6.00005-6

Table 5.1 Aerodynamic model versions overview

Model name: Aero-model A *Main assumptions*: Blade element theory (BET)—coarse BE grid, model assumptions are adjusted for clamped conditions Number of blade elements: 3 Input kinematics: Vicon Vision—E1 bench test (ML101, MSK004) *Model code:* Aero-model A load experimental bench test (MATLAB: AEROALEB.m)
Model name: Aero-model B *Assumptions*: BET medium–refined BET grid, model assumptions are for free-flight conditions Number of blade elements: 8 Input kinematics: Vicon Vision ®—E2 Flight Test (ML101) *Model code*: Aero-model B load experimental free flight (MATLAB: AEROBLEF.m)
Model name: Aero-model C *Assumptions*: BET-refined BET grid, model assumptions are for free-flight conditions, input kinematics formulated in terms of generalized coordinates of a five-body flexible multi-body dynamics model of ornithopter, dynamic stall included Number of elements: 13 *Model code*: Aero load experimental free flight (MATLAB: AEROLEF.m) Input kinematics: Vicon Vision—E2 flight test (ML101) *Model code*: Aero load initialized experimental coupled (MATLAB:AEROIEC.m) Input kinematics: Output of five-body dynamics model, initialization is through Vicon Vision—E2 flight test (ML101)

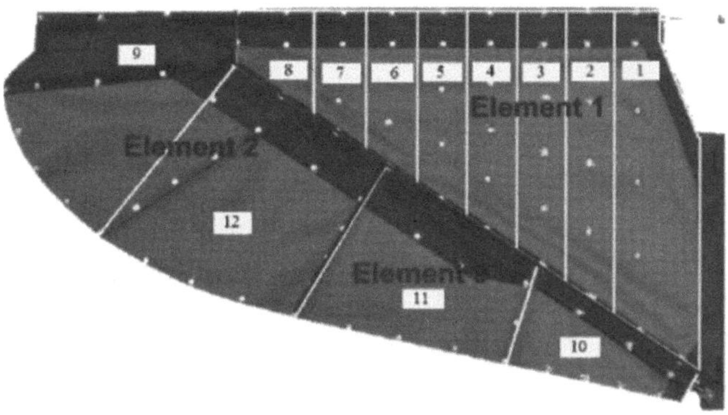

Figure 5.1 Blade element grid schematic aero-model A.

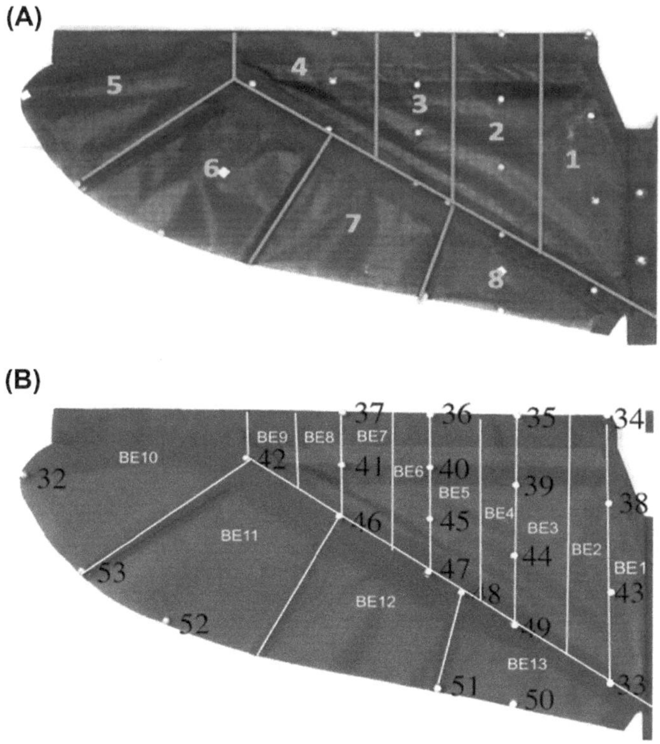

Figure 5.2 (A) Blade element grid schematic aero-model B. (B) Blade element (BE) grid schematic aero-model C.

used to correlate and improve model C, implemented in AEROLEF. The final and improved model C was further developed to be coupled with the FMBDM as described in the next section. AEROIEC, code of aero-model C ties around the five-body vehicle dynamics model.

Aero-model A served mainly as a proof of concept study for the integration with a FMBDM. It used wing kinematics data from experiment E1 in clamped conditions for the aerodynamic force generation and did not account for circulatory effects. Aerodynamic model B utilized wing kinematics data in free-flight E2, a refined blade element selection, and accounted for circulatory and noncirculatory effects. Blade element selection was further refined in model C, where there was a 35% improvement in correlation of integrated forces to experimental data E2. Like model B, final aerodynamic model C also utilized free-flight test kinematics E2 with a further refined blade element selection. Model B was further improved to aerodynamic model C, mainly by the inclusion of dynamic stall assumptions. Implementation of aerodynamic model C is described in detail in this chapter. That is the workflow of the model C. Implementation of earlier model iterations of model A and B is analogous. For models A, B, and C, assumptions used and improvements between the models versions are also discussed in the reminder of this chapter.

5.2 AERO-MODEL A

Harmon's blade element theory was used for time-variant forces on the wing [1]. Ornithopter test platform used for validation and verification was clamped to a ground stand. Therefore, in the iteration of this model, only time-variant noncirculatory aerodynamic forces were applied to the wing structural model. It consists of 1475 nodes. Aerodynamic loads were applied to the correspondent nodes in the wing region as defined in its blade element selection. Experimental wing kinematics (E1) of the research ornithopter (ML101) provided local wing deformation during a flapping cycle for the aerodynamic force calculation. Time-variant forcing functions were applied to corresponding luff and thrust flap region shown in Fig. 5.1.

Owing to zero forward velocity at the clamped conditions simulated, it was assumed that only noncirculatory aerodynamic loads were acting on the wing. The noncirculatory force is the result of the acceleration of a body of air normal to the wing surface as the wing undergoes quasi-steady motion. Local circulatory lift force acts normal to the inflow velocity, producing both vertical and horizontal forces. For a flapping and pitching wing with a nonzero flight speed, the noncirculatory normal force generated by the wing is given by (Eq. 5.1) [1].

$$dN_{nc} = -\frac{\rho\pi\left(dc^2\right)}{4}\left(\dot{\theta}U + r\ddot{\beta}\cos\theta - ba\ddot{\theta}\right)dr \tag{5.1}$$

This is the normal force value for a wing segment of length dr, where r is the mean spanwise position of the segment and dc is the chord length. Here a is the location of airfoil pitching access from 0.5 chord point, b is the reference span, and c is the reference chord. The local angular acceleration of the flapping wing is denoted by $\ddot{\beta}$, and θ is the pitch angle of blade element with respect to flapping axis.

The noncirculatory force acts normal to the wing, and therefore is broken down into vertical and horizontal components as well. Since the local blade element orientation varies along the span, the vertical and horizontal components of each force were computed for the blade element forces. The resulting vertical and horizontal components of the calculated noncirculatory forces are according to Eqs. (5.2) and (5.3), respectively.

$$dF_{vert\_nc} = dN_{nc}\cos(-\theta)\cos(\partial + \gamma)\cos\beta \tag{5.2}$$

$$dF_{horiz\_nc} = dN_{nc}\sin(-\theta)\cos(\partial + \gamma) \tag{5.3}$$

Here ∂ is the angle between flapping axis and flight velocity, γ the flight path angle of flight velocity with respect to the ground and ϕ is the pitch angle of blade element chord with respect to flapping axis. Kinematic data for the calculations are provided by experimental data. For the sum of element forces over each wing region, see in Fig. 5.1. For the simulations, summed time-variant forces are applied on the luff and the thrust flap region accordingly. This is assuming uniform special load distribution of the time-variant loads at each wing region. Time-variant forcing functions resulting from previously mentioned blade element theory were applied in the form of linear combinations of sign waves. Further details about used aerodynamic modeling methodology can be found in Ref. [19]. Aerodynamic loads and resulting mode shapes of the wing were computed for the flapping frequency of 6.17 Hz and compared to experiment in Section 6.3.1 to assess the quality of the model.

5.3 AERO-MODEL B/C

For the aerodynamic model C assumptions according to Lee have been adopted, which is an improved DeLaurier model. These assumptions account for dynamic stall cut off, chordwise wing flexibility, and introduce an attached flow condition, a dynamic stall condition, and a poststall condition [4,5]. Model C furthermore accounts for flexibility in the span and additional flexibility in chordwise direction due to a fore-and-aft approach of the wing sections. This represents an improvement in aero-model C compared to previously used assumptions. An improved fluid—structure interaction (FSI) procedure, where aerodynamic forces are formulated as a continuous function and estimated as a superposition of sign waves, lead to model reduction. This time variant forces in form of the continuous function lead to a more efficient FSI model. Input kinematics were based on inflight kinematics (E2) for the model development and later formed the first set of generalized input coordinates for the coupled model. This lead to an earlier convergence of a solution based on real inflight dynamics.

5.3.1 Aerodynamic Model

Improvements to aero-model C compared to aero-model B include dynamic stall and a blade element refinement according to Fig. 5.2A and B. The blade element selections and equations for the aerodynamic force calculations for aero-model C are further discussed. Aero-model B is analogous and excludes dynamic stall cut off, the attached flow condition, a dynamic stall condition, and a poststall condition compared to aero-model C. For the free-flight aerodynamics models B and C, the local airspeed with respect to the vehicle is given by the vector $v(\xi_{IB})$ that is the free speed velocity of the vehicle

$$v(\xi_{IB}) = \sqrt{\dot{x}_{IB}{}^2 + \dot{y}_{IB}{}^2 + \dot{z}_{IB}{}^2} \tag{5.4}$$

where $\dot{x}_{IB}{}^2$, $\dot{y}_{IB}{}^2$, and $\dot{z}_{IB}{}^2$ are the translational body fixed velocity components on CB0. Fig. 4.2 shows the location of the origin of the wing root fixed reference frame CB0. The flapping axis of the wing is orientated by the pitch angle $\theta = \theta_{IB} + \theta_{IM1}$ to the velocity vector $v(\xi_{IB}) = U$ in the luff region, and $\theta = \theta_{IB} + \theta_{IM1} + \sqrt{\psi_{IT}^2 + \theta_{IT}^2}$ in the thrust flap region.

High angles of attack occur in flight [refer to flight test data (E2)]. Such bioinspired flight is accounted for by including higher-order approximations in the aerodynamics model. For variables used for the aerodynamic force calculation described next, see Fig. 5.3.

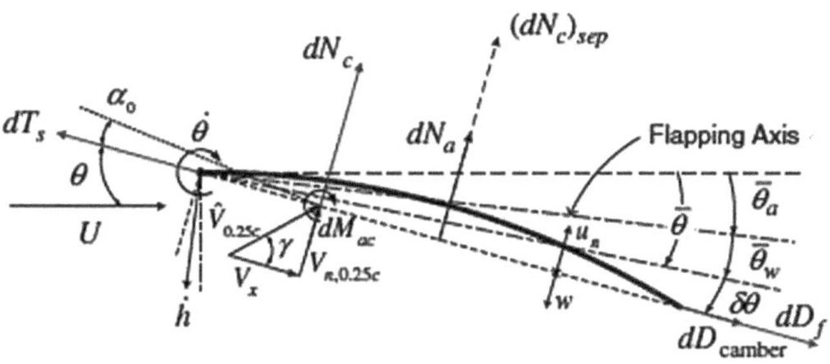

Figure 5.3 Variables for calculation of aerodynamic forces [1].

Variables are defined for the calculation of the circulatory normal force, the leading edge suction force, and the chordwise force due to the camber [1]. Note that these take the high relative angle of attack into account. The other aerodynamic forces, dN_a and dD_f, according to the DeLaurier model are also described here [2].

The total attached flow normal force is [1].

$$dN = dN_c + dN_a \tag{5.5}$$

where dN_a is the noncirculatory normal force and dN_c the circulatory normal force. The normal force contribution from the apparent mass effect acts at the mid-chord is as follows [2];

$$dN_a = \frac{\rho\pi c^2}{4}\ddot{z}_2 dy \tag{5.6}$$

where y is the coordinate along the semispan defined in Fig. 4.2; $\ddot{z}_2$ is the time rate of change of the mid-chord normal velocity component due to the wings motion. The circulatory normal force used is according to [1],

$$dN_c = 2\pi\left(\alpha' + \alpha_0 + \overline{\theta}\right)\cos\gamma\frac{\rho U V_{0.25c}{}^2}{4}cdy \tag{5.7}$$

where $V_{0.25c}$ is the resultant velocity at the quarter-chord location, γ the relative AOA. $\overline{\theta}$ is the total mean pitch angle, α_0 is the zero lift AOA, and α' the flow's relative angle of attack at the 0.75 chord location. The leading edge suction force was accounted for using following equation [1],

$$dT_s = \eta_s 2\pi\left(\alpha' + \overline{\theta} - \frac{1c\dot{\theta}}{4U}\right)\sin\gamma\frac{\rho U \widehat{V}_{0.25c}{}^2}{4}cdy \tag{5.8}$$

here η_s is the leading edge suction efficiency and $\dot{\theta}$ is the pitch velocity of the blade element. The chordwise force due to the camber was calculated in the model using [1],

$$dD_{camber} = -2\pi\alpha_0\left(\alpha' + \overline{\theta}\right)\cos\gamma\frac{\rho U \widehat{V}_{0.25c}^{\ 2}}{2}cdy \tag{5.9}$$

and the chord-friction drag was used as follows [2],

$$dD_f = (C_d)\frac{\rho V_x^{\ 2}}{2}cdy \tag{5.10}$$

here V_x is the flow speed tangential to the section. And V itself is the flow velocity, which includes the downwash as well as the wing's motion relative to U [2]. Vertical velocity V_x at the quarter-chord location is according to Eq. (5.3) in [2].

The minimum and maximum dynamic stall angles according to Eqs. (5.9)−(5.11) in [1] depend on plunging and pitching effects and are included in the model.

Aerodynamic force generation motion variables in the above equations were formulated in terms of the generalized coordinates of the five-body flexible dynamics model for the model implementation, to couple aero-model C to the five-body flexible model and is described in the next section.

Time-variant aerodynamic forces resulting from equations in this section were applied onto the nodes of the corresponding blade elements of the wing model. For the special distribution of the blade element section, an elliptical shape was assumed. In the remainder of this section the aero-load blade element selection is discussed.

5.3.2 Aero-Load Blade Element Identification

Owing to significant thrust flap deflections, which are much larger than the small flap angle, the assumption used in the thin airfoil theory, a fore-and-aft blade element approach was used in this analysis in terms of generalized coordinates of the system. This leads to more accurate results versus covering the entire wing length with blades [19]. The trailing edge flap is a significant portion of the length and made this approach favorable. See Harmon and Hubbard [1] for more details. Blade element selection for model C was used according to Table 5.1.

The free-flight test analysis showed that only out-of-plane flexibility is significant (Section 6.1). Wing flexibility and camber forming leads to insignificant shortening of blade element width dr. Therefore, the blade element width dr was assumed to remain constant throughout the computation. Time-independent variables y and x used in Table 5.2 refer to marker coordinates within its reference frame in PS0 position as discussed in Section 6.1.1.1.

The maximal blade element chord length is c_{max} and also computed from the free-flight data where the maximal element length during the flapping cycle was used for the variable (according to PS0). For specifics about blade element length and calculation, see Table 5.2. The blade element identification during a flapping cycle is according to Fig. 5.4.

The blade element lengths for elements 1−9 are identified with the marker position on their outer edge with an adjustment for the length in the middle of the blade element. These markers lying between the outer marker points of a blade element identify the local camber. See Blade Element 3 according to Fig. 5.6. Markers 12/17/22 and 10 are used to compute the local camber. Blade elements 10−13 are calculated as the diagonal spar distance of markers bordering the blade element. See blade element 11 mirrored marker 42 according to Fig. 5.4. Blade element 10 is of irregular shape, and the blade element with dr_{10} and length dc_{10} shown in Fig. 5.5 provides a good approximation. Mainly the blade element width and length are utilized to calculate the surface area.

The camber-line (or chord-line) data points recorded for all blades and a set of matrix coordinates are representing the blade element position states to the programs, "AEROLEF.m and AEROIEC.m". The matrix coordinates are within the reference frame of their global location and orientation in the body reference frame system.

5.4 THE AERODYNAMIC MODEL IMPLEMENTATION

The workflow of aero-model C is described in the following. Experimental wing kinematics were formulated in terms of the generalized coordinates of the five-body flexible dynamics system as shown in Fig. 5.6. The aerodynamic model was used to compute time-variant forces on the wing.

The experimental wing deflections E2 were measured in the inertia reference frame using the Vicon Vision, as shown in Fig. 5.6.

The experimental data were recorded using 34 time steps ($ti = 34$) over one flapping cycle. Measurements were first transformed into the local body fixed reference frame, CB, then to the wing fixed reference frames, CW and

Table 5.2 Blade element selection ML101 ornithopter configuration

Number of blade element n	Blade element chord length dc_n	Blade element with dr_n
$n = 1$	$dc_1(t) = \left(x(t)_{M33}\right) + sl_d\left(\frac{y_{M33}}{2}\right)$	$dr_1 = y_{M33}$
$n = 2$	$dc_2(t) = \left(x(t)_{M10}\right) + sl_d\left(\frac{y_{M10} - y_{M33}}{2}\right)$	$dr_2 = y_{M10} - y_{M33}$
$n = 3$	$dc_3(t) = \left(x(t)_{M49}\right) + sl_d\left(\frac{y_{M49} - y_{M10}}{2}\right)$	$dr_3 = y_{M49} - y_{M10}$
$n = 4$	$dc_4(t) = \left(x(t)_{M29}\right) + sl_d\left(\frac{y_{M29} - y_{M49}}{2}\right)$	$dr_4 = y_{M29} - y_{M49}$
$n = 5$	$dc_5(t) = \left(x(t)_{M47}\right) + sl_d\left(\frac{y_{M47} - y_{M29}}{2}\right)$	$dr_5 = y_{M47} - y_{M29}$
$n = 6$	$dc_6(t) = \left(x(t)_{M28}\right) + sl_d\left(\frac{y_{M46} - y_{M47}}{4}\right)$	$dr_6 = \frac{y_{M46} - y_{M47}}{2}$
$n = 7$	$dc_7(t) = \left(x(t)_{M46}\right) + sl_d\left(\frac{y_{M46} - y_{M47}}{4}\right)$	$dr_7 = \frac{y_{M46} - y_{M47}}{2}$
$n = 8$	$dc_8(t) = \left(x(t)_{M25}\right) + sl_d\left(\frac{y_{M25} - y_{M64}}{2}\right)$	$dr_8 = y_{M25} - y_{M64}$
$n = 9$	$dc_9(t) = \left(x(t)_{M42}\right) + sl_d\left(\frac{y_{M42} - y_{M25}}{2}\right)$	$dr_9 = y_{M42} - y_{M25}$
$n = 10$	$dc_{10}(t) = \sqrt{\left(x(t)_{M53} - x(t)_{M16}\right)^2 + \left(y_{M53} - y_{M16}\right)^2}$	$dr_{10} = \sqrt{\left(x_{M42}\right)^2 + \left(y_{M9} - y_{M42}\right)^2}$
$n = 11$	$dc_{11}(t) = \sqrt{\left(x(t)_{M52} - x(t)_{M25}\right)^2 + \left(y_{M55} - y_{M25}\right)^2}$	$dr_{11} = \sqrt{\left(x_{M27} - x_{M25}\right)^2 + \left(y_{M25} - y_{M27}\right)^2}$
$n = 12$	$dc(t)_{12} = \sqrt{\left(x(t)_{M31} - x(t)_{M28}\right)^2 + \left(y_{M31} - y_{M28}\right)^2}$	$dr_{12} = \sqrt{\left(x_{M48} - x_{M27}\right)^2 + \left(y_{M27} - y_{M48}\right)^2}$
$n = 13$	$dc_{13}(t) = \sqrt{\left(x(t)_{M50} - x(t)_{M10}\right)^2 + \left(y_{M50} - y_{M10}\right)^2}$	$dr_{13} = \sqrt{\left(\left(x_{M33} + sl_d\left(y_{M33} - y_2\right)\right) - x_{M48}\right)^2 + \left(y_{M48}\right)^2\left(-x_{M48}\right)^2 + \left(y_{M48}\right)^2}$

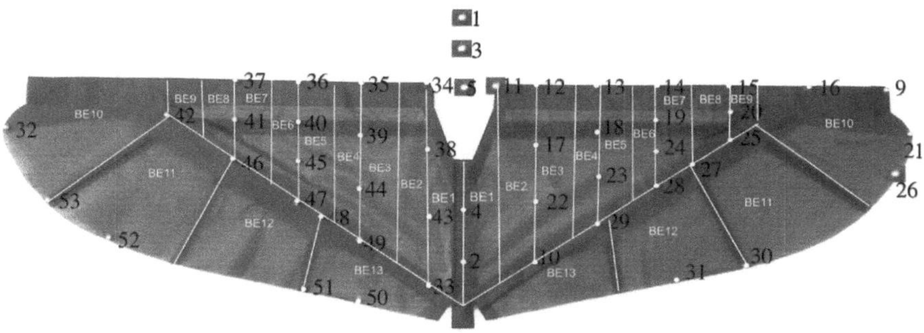

Figure 5.4 ML101 ornithopter configuration blade element (BE) selection.

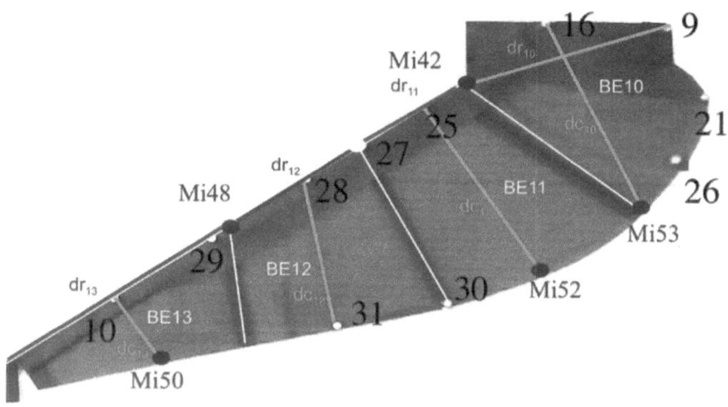

Figure 5.5 Blade element (BE) selection thrust flap region ML101 ornithopter configuration.

CT. See Fig. 4.3. The underlying reference motion of the five-body dynamics system was then removed. The free-flight velocity data from the flight test E2 are of the vehicle fuselage fixed reference frame, $\nu(\xi_{IB(E)})$, relative to the inertial reference frame. This serves as an input into the aerodynamic model C. The generalized fuselage coordinates are:

$$\xi_{IB(E)} = \left\{ \begin{matrix} x_{IB} \\ y_{IB} \\ z_{IB} \\ \psi_{IB} \\ \theta_{IB} \\ \phi_{IB} \end{matrix} \right\}_{ti=34} \tag{5.11}$$

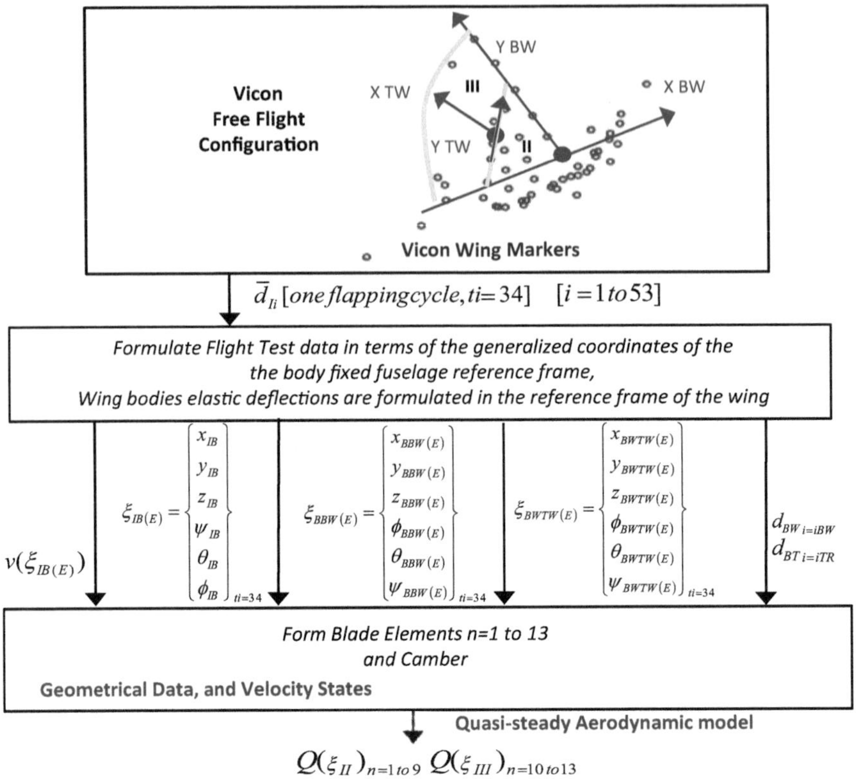

Figure 5.6 Aerodynamic loads model—workflow. FMBD, flexible multi-body dynamics.

See Fig. 4.6 for nomenclature reference. The generalized coordinate of the thrust flap region,

$$\xi_{BBW(E)} = \begin{Bmatrix} x_{BBW(E)} \\ y_{BBW(E)} \\ z_{BBW(E)} \\ \phi_{BBW(E)} \\ \theta_{BBW(E)} \\ \psi_{BBW(E)} \end{Bmatrix}_{ti=34} \tag{5.12}$$

The generalized coordinate of the thrust flap region,

$$\xi_{BWTW(E)} = \begin{Bmatrix} x_{BWTW(E)} \\ y_{BWTW(E)} \\ z_{BWTW(E)} \\ \phi_{BWTW(E)} \\ \theta_{BWTW(E)} \\ \psi_{BWTW(E)} \end{Bmatrix}_{ti=34} \tag{5.13}$$

Eqs. (5.1)–(5.21) define the coordinates of the underlying rigid-body motion of the wing. The coordinate matrix of the deformation emanating from the underlying rigid-body motion is d_i and is the deformation vector of each marker location on the wing fixed reference frame. In luff, the deformation vector is $d_{BWi=iBW}$ and $d_{BTi=iTR}$ in the thrust flap region. The markers comprising the coordinate matrix over a flapping cycle in the thrust flap region are $iBT=$ Wing Markers 9, 16, 21, 26, 30, 31, 32, 50–53, and the markers in the wing region are $iBW=$ Wing Markers 2, 4, 5, 9–53. The deformation (*di*) matrix is in time steps (*ti*) and described in the generalized coordinates of the region. The blade element formation uses the coordinate matrix in this form to form blade elements, calculate camber geometry and translational and angular velocities needed for the aerodynamic load calculations on the wing. See previous section for aerodynamic load calculation and assumptions.

The aerodynamic model load output results in a generalized external force in the multi-body dynamics model equations of motion (EOMs). For each blade element for the two wing bodies [Body II (blade elements $n = 1$ to 9) and Body III (blade elements $n = 10$ to 13)], the aerodynamic load appears in the generalized loads term $Q(\xi_{II})_{n=1 \text{ to } 9}Q(\xi_{III})_{n=10 \text{ to } 13}$ in the EOM of the five-body vehicle dynamics system. $Q(\xi_{II})_{n=1 \text{ to } 9}$ and $Q(\xi_{II})_{n=1 \text{ to } 9}$ are further discussed in the next section.

d_i is the deformation vector of each marker location on the wing fixed reference frame (experimental data) and are $\mathbf{r}_{PP'}$ (as defined in Chapter 4) deformation of the marker in the five-body vehicle dynamics model. Recall $\mathbf{r}_{PP'}$ is the physical coordinate vector from the undeformed position of a marker P on the wing to its deformed position P' and can be obtained from the modal coordinates, $\mathbf{q}$ in the wing reference frame. Experimental kinematics (subscript E in the following) are interchangeable with five-body flexible vehicle dynamics model coordinates in the aerodynamic model. This is due to the formulation of the aero-model C input data in terms of the generalized coordinates of the five-body vehicle dynamics system.

In AEROIEC generalized coordinates of the free-flight test, data are replaced by kinematics resulting from the model Fig. 5.7.

Experimental free-flight test data serves as an input variable for the aero-model C initialization in AEROIEC. Output forces were projected into the generalized coordinate space and applied to the FMBDM. The code using model C solves the generalized coordinates in the system, the generalized coordinates definition was discussed in Chapter 4. These

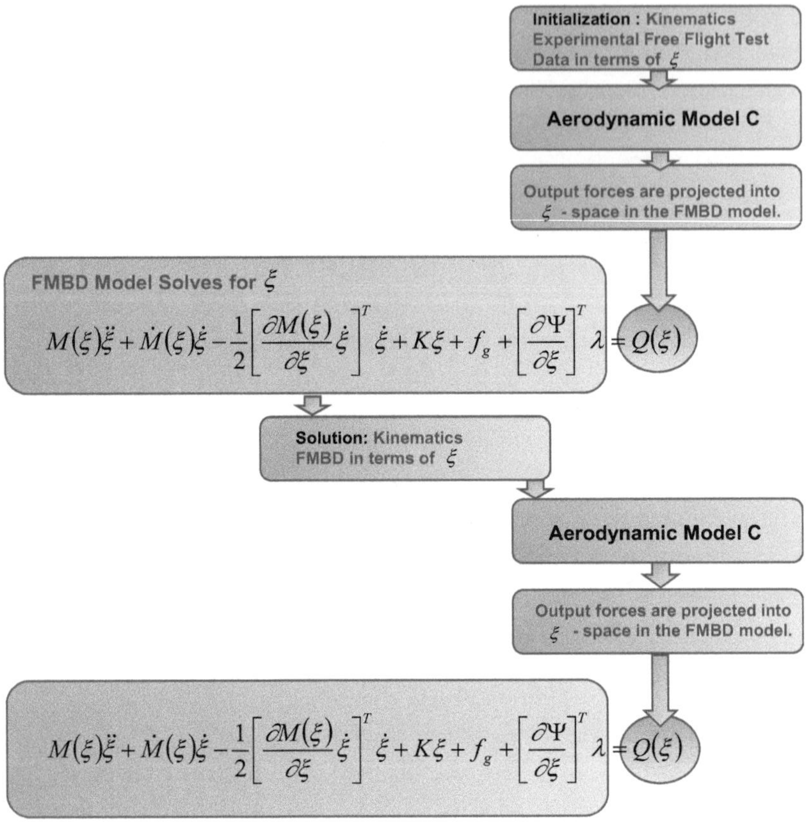

Figure 5.7 Aerodynamic loads model—workflow Aero Load Initialized Experimental Coupled. *FMBD*, flexible multi-body dynamics.

previously described position states of the generalized coordinates now serve as input into aero-model C and result in aerodynamic loads, which now serve as input into FMBDM EOMs. Therefore, different configurations in underlying wing motions and wing stiffnesses can be run within the integrated FMBDM simulation using AEROIEC, and a solution can be correlated.

5.4.1 Distributed Aerodynamic Loads

Distributed aerodynamic loads resulting from aero-model C were introduced to the five-body vehicle dynamics model using an MFORCE element [1]. The distributed load vector was applied to the finite element method (FEM) elements in the model using a half-elliptical special distribution on the blade [1].

The physical coordinate from the EOMs in the FEM software were examined and discussed:

$$M\ddot{x} + Kx = F \tag{5.14}$$

K and M are the finite element mass and stiffness matrices for the flexible component, respectively; and x and F are the physical nodal degree of freedom (DOF) vector and the load vector, respectively [3].

Now one transforms into modal coordinates q using the modal matrix $\mathbf{\Phi}$ (Eq. 5.14) becomes,

$$\mathbf{\Phi}^T M \mathbf{\Phi} \ddot{q} + \mathbf{\Phi}^T K \mathbf{\Phi} q = \mathbf{\Phi}^T F \tag{5.15}$$

which simplifies to the following EOM.

$$\widehat{M}\ddot{q} + \widehat{K}q = f \tag{5.16}$$

where $\widehat{M}$ and $\widehat{K}$ are the generalized mass and stiffness matrices, respectively and f is the modal load vector [3].

The applied aerodynamic force is having a global resultant force and moment. These show up as loads on the rigid-body modes and treated as point forces and moment on the local reference frame.

The projection of the nodal force vector on the modal coordinates [3]

$$f = \mathbf{\Phi}^T F \tag{5.17}$$

is a computationally expensive operation, which poses a problem when F is an arbitrary function of time and is the case for the time-variant aerodynamic force [3].

The force is now separated in space and time and the load can be viewed as a time-varying linear combination of an arbitrary number of static load cases according to following:

$$F(t) = s_1(t)F_1 + \cdots + s_n(t)F_n \tag{5.18}$$

Therefore, the expensive projection of the load to modal coordinates is performed once during the creation of the MNF file according to Section 4.7.1. The multi-body dynamics calculation performed by the Adams solver only needs to be aware of the modal form of the load according to the following:

$$f(t) = s_1(t)f_1 + \cdots + s_n(t)f_n \tag{5.19}$$

where the vectors f_1 to f_n are n different case vectors. Each of the load case vectors contains one entry for each mode in the modal basis [1].

One can reformulate to have an explicit dependency on the system response, where q represents all the states of the system. The modal force of the system can now be described according to Eq. (5.20)

$$f(q,t) = s_1(q,t)f_1 + \cdots + s_n(q,t)f_n \tag{5.20}$$

The applied aerodynamic force has a global resultant force and moments. These show up as loads on the rigid-body modes and are treated the multi-body dynamics model as point forces and moments on the local wing fixed reference frames.

5.5 THE GLOBAL RESULTING FORCES AND MOMENTS

A point force F and a point moments ME are applied to a marker location on a flexible body. They must be projected on the generalized coordinates of the flexible multi-body system.

The force and torque formulated in matrix form. They are expressed in the coordinate system of the marker L in the local reference frames of the wing bodies ZBW and ZBT (Fig. 4.3).

$$F_L = \begin{bmatrix} f_x \\ f_y \\ f_z \end{bmatrix}, \quad ME_L = \begin{bmatrix} me_x \\ me_y \\ me_z \end{bmatrix} \tag{5.21}$$

With the number of markers L corresponding to the number of nodal points, hence $L = 1$ to nl, where nl denotes the number of nodes a force is applied on Bodies II (L/R), and $L = 1$ to nm, where nm is the number of nodes on Bodies III (L/R) where a force is applied.

5.5.1 Generalized Force Definition

The generalized force Q consists of a generalized translational force Q_F, a generalized moments Q_{ME}, which is the generalized force on the Euler angles and generalized modal force Q_M. Therefore it follows,

$$Q = \begin{bmatrix} Q_F \\ Q_{ME} \\ Q_M \end{bmatrix} \tag{5.22}$$

For the EOM in the inertia reference frame the generalized translational force is obtained by a coordinate transformation for the Q_F of Body II and Body III. Q_F for the Bodies II and III follows respectively,

$$Q^I{}_{FII} = {}^I A^{LII} F_{LII} \tag{5.23}$$

$$Q^I{}_{FIII} = {}^I A^{LIII} F_{LIII} \tag{5.24}$$

The transformation matrix ${}^I A^{LII}$ is given in Eq. (63) and the transformation matrix ${}^I A^{LIII}$ in Eq. (64) in Chapter 4, Section 4.6.3.

For the total integrated forces in the inertia reference frame due to applied aerodynamics loads, it follows the summation of applied loads on Wing Bodies II (L/R) and Wing Bodies III (L/R).

$$Q^I{}_F = 2x\left(\sum_{l}^{nl} Q^I{}_{FII} + \sum_{l}^{nm} Q^I{}_{FIII}\right) \tag{5.25}$$

These applied forces vary per blade element and are further described in Section 5.4.

For verification, the total integrated, translational aerodynamic forces obtained the fuselage fixed reference frame CB according to the following:

$$Q^{CB}{}_{FIIL} = {}^{CB} A^{LII} F_{LII} \tag{5.26}$$

$$Q^{CB}{}_{FIIIL} = {}^{CB} A^{LIII} F_{LIII} \tag{5.27}$$

The transformation matrices are according to Eqs. (5.28) and (5.29), respectively. The coordinate transformation is analogous to description in Section 4.6.3.

$${}^B A^{LIII} = {}^B A^{M1\,M1} A^{T\,T} A^{P\,P} A^{LIII} \tag{5.28}$$

$${}^B A^{LII} = {}^B A^{M1\,M1} A^{P\,P} A^{LII} \tag{5.29}$$

For the combined forces on the bodies, it is the summation of the individual forces over the nodes, and it follows respectively for the combined aerodynamic forces on a wing body.

$$Q^{CB}{}_{FII} = \sum_{n=1}^{nl} Q^{CB}{}_{FIIL^{nl}} \tag{5.30}$$

$$Q^{CB}{}_{FIII} = \sum_{n=1}^{nm} Q^{CB}{}_{FIIIL^{nm}} \tag{5.31}$$

Total forces and moments are also computed in the fuselage fixed body reference frame CB according to following summation, $Q^{CB}{}_{FI}$ denotes the generalized translational force applied on the fuselage body.

$$Q^{CB}{}_{FT} = 2x\left(Q^{CB}{}_{FII} + Q^{CB}{}_{FIII}\right) + Q^{CB}{}_{FI} \tag{5.32}$$

The generalized force due to the aerodynamic loads applied on the wings is

$$Q^{CB}{}_{FT} = 2x\left(Q^{CB}{}_{FII} + Q^{CB}{}_{FIII}\right) \tag{5.33}$$

5.5.2 Generalized Moment Definition

The generalized moments Q_{ME} on the flexible wings are due to the force vector F_L and the moment vector ME_L. It follows for the generalized moments of Bodies II (L/R).

$$Q_{MEII} = \sum_{nl=1}^{nl} ME_{LII} + \sum_{nl=1}^{nl} \overrightarrow{p}_{B0A^{nl}} x F_{LII^{nl}} \tag{5.34}$$

where the vector $\overrightarrow{p}_{B0A^{n}}$ is the position vector form, the origin of the local body reference frame CBW ($B0$) of Bodies II (L/R) to the nodal point nl of force application. For Bodies III (L/R) it follows

$$Q_{MEIII} = \sum_{nm=1}^{nm} ME_{LIII} + \sum_{nm=1}^{mn} \overrightarrow{p}_{TA^{nl}} x F_{LIII^{nm}} \tag{5.35}$$

The vector $\overrightarrow{p}_{TA^{nm}}$ is the position vector from the origin of the local body reference frames CBT(T) to the nodal point's nm of force application on Bodies III (L/R).

The total generalized moments Q_{ME} in the fuselage fixed reference frame are computed according to following:

$$Q_{ME} = 2x\left(Q_{MEII} + Q_{MEIII}\right) + Q_{MEI} \tag{5.36}$$

5.5.3 Generalized Modal Force Definition

The generalized modal force calculation on the flexible bodies is described in this section. The forces are transformed in the local wing fixed reference frame of the flexible bodies (II/III—L/R). The generalized modal force results from the multiplication with the corresponding slice of the modal matrix.

The generalized modal force of bodies II (L/R) is according to the following:

$$Q_{MII} = \Phi_P^T Q_{FII}^{CB} + \Phi_P^{*T} Q_{MEII}^{CB} \tag{5.37}$$

For bodies III (L/R) according to

$$Q_{MIII} = \Phi_P^T Q_{FIII}^{CB} + \Phi_P^{*T} Q_{MEIII}^{CB} \tag{5.38}$$

where Φ_P^T and Φ_P^{*T} are the slices of the modal matrix corresponding to the translational and angular DOF discussed earlier in this chapter.

The total modal force due to the aerodynamic forces observed in the fuselage body fixed reference frame becomes

$$Q_M = 2 \times (Q_{MII} + Q_{MIII}) \tag{5.39}$$

From the aerodynamic loads resulting in generalized translational force Q_F, generalized moments Q_{ME} and a generalized modal force Q_M augment the generalized force vector Q (Eq. 5.22). Torque on the wing root due to motor torque additionally augments $Q(\xi)$ in EOM.

REFERENCES

[1] R.L. Harmon, J.E. Hubbard, Aerodynamic Modeling of a Flapping Membrane Wing Using Motion Tracking Experiments (Master's thesis), 2008.

[2] D. DeLaurier, An aerodynamic model for flapping-wing flight, The Aeronautical Journal 97 (964) (April 1993) 125–130.

[3] MSC Software, Adams Flex – Reported in Adams Docs DOC10403, MSC, 2013.

[4] J. Lee, D. Kim, J. Han, Cycle-averaged mean and zero-mean time-varying flight dynamics of the ornithopters at trim, Smart Systems and Structures Lab, KAIST (2010).

[5] D. Kim, S. Lee, J. Han, Improved aerodynamic model for efficient analysis of flapping-wing flight, AIAA Technical Journal 49 (4) (2011).

CHAPTER SIX

Results of the Modeling Methodology Implementation and Flight Simulation

6.1 MODELING ASSUMPTIONS VERIFICATION AND WING FLEXIBILITY

6.1.1 Coordinate Variations

To check modeling assumptions and inherent wing flexibility in the experimental flight test data set, E-2 was first computed and analyzed in the fuselage body fixed reference frame CB0 that is located on the wing root. The experimental resulting position states of all the tracking markers $i = 1-53$ at the verification frequency of 6.06 Hz are shown in Figs. 6.1–6.3.

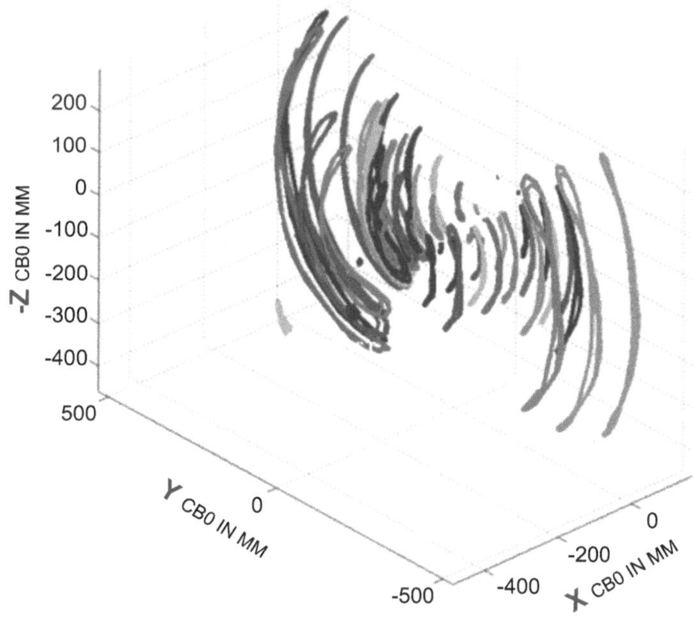

Figure 6.1 Isometric view of the position states path of all Vicon markers over one flapping cycle at 6.06 Hz.

Modern Flexible Multi-Body Dynamics Modeling Methodology for Flapping Wing Vehicles
ISBN: 978-0-12-814136-6
http://dx.doi.org/10.1016/B978-0-12-814136-6.00006-8

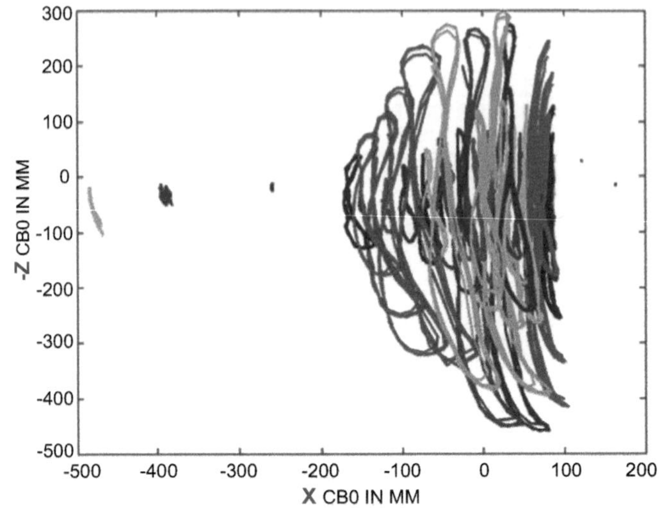

Figure 6.2 Side view of the position states path of all Vicon markers over one flapping cycle at 6.06 Hz.

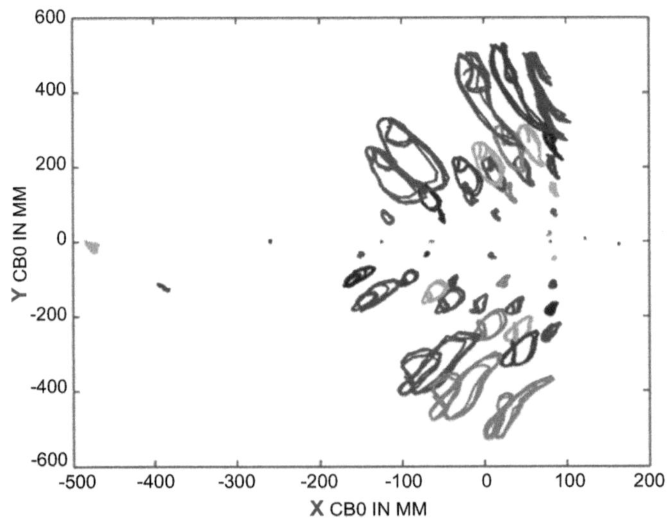

Figure 6.3 Top view of the position states path of all Vicon markers over one flapping cycle at 6.06 Hz.

6.1.1.1 Zero Deflection Reference Plane

Fig. 6.4 shows the zero deflection reference plane (PS0) of tracking markers in the flight test data used for referencing purposes in the following analysis. The position state is named PS0. The zero deflection reference plane is

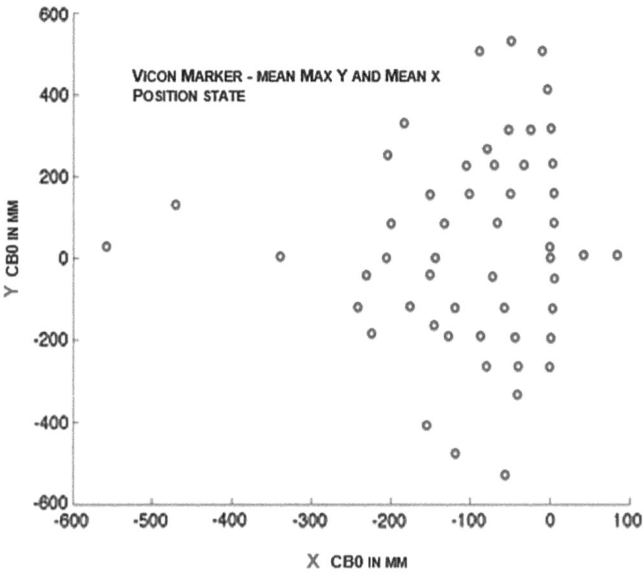

Figure 6.4 Top view of the ornithopter markers position state 0 deflection plane.

associated with a zero flapping angle (FA) position. These represent the wing marker locations in free flight in an extended position that is the maximum norm of the x/y position in CB0 (over a flapping cycle). It represents the x/y plane of the wing fixed reference plane CW defined in Chapter 4.6. For PS0 the mean maximum norm of y and mean maximum norm of x of $i = 1-53$ was computed over 125 position states.

6.1.1.2 Quantification Wing Flexibility

To estimate the wing flexibility in y, the x location is maintained as a constant according to the definition of PS0. Think of PS0 as the marker locations of a rigid wing configuration.

Analyzing the flight test data E-2 in all position states during the flapping cycle results in a mean standard deviation (STD) of 18.054 mm in the y location (yCW) of all marker points on the wing. This indicates the y deflection and therefore flexibility in the wing fixed reference coordinate system CW during the flapping cycle. Sample size is 125 position states, representing 3.5 flapping cycles, and averaged and plotted for one flapping cycle in Fig. 6.5.

The wing flexibility in x computed analogous to y resulted in a mean STD of 7.12 mm in x (xCW) of all marker points and position states during one flapping cycle (Fig. 6.6).

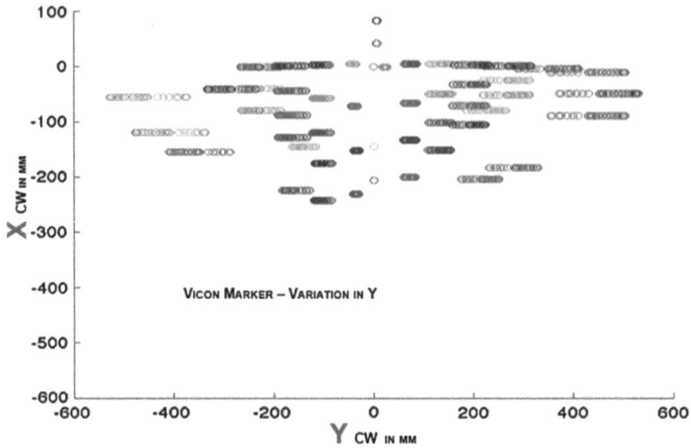

Figure 6.5 Top view of the *Y* coordinate flexibility in the wing fixed reference plane CW.

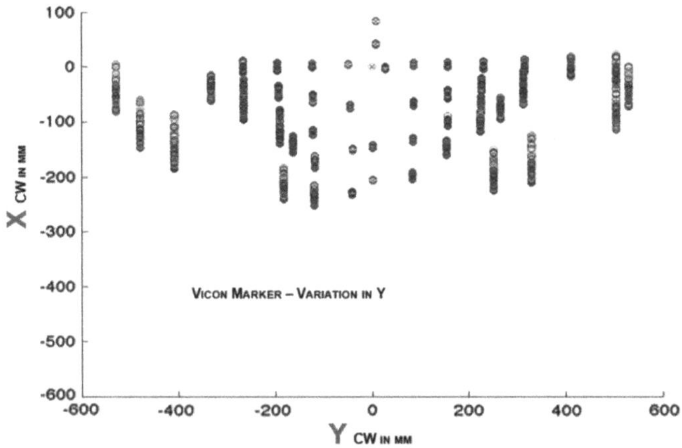

Figure 6.6 Top view of the *X* coordinate flexibility in the wing fixed reference plane CW.

In summary, a mean deformation in x of all tracking markers on the wing of 7.12 mm and a mean deformation in y of 18.05 mm in the wing fixed reference frame CW found during a flapping cycle defined the flexibility of the experimental wing ML101 in free flight.

6.1.2 Large Global Deformations

All position states during a flapping cycle of the wing markers of experimental data E-2 were mirred around YCB0 to obtain a refinement of wing markers on one wing. By computing a polynomial fit between marker locations in all position states during a flapping cycle, the wing surface of the

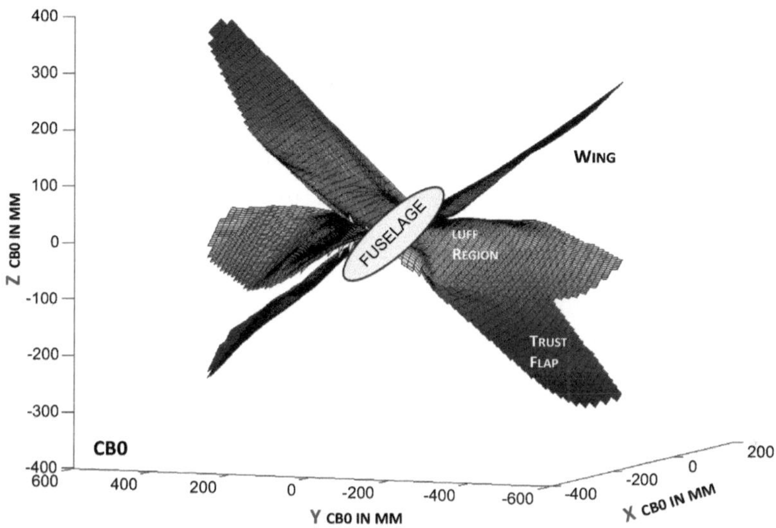

Figure 6.7 Bioinspired ornithopter test platform in free flight: three experimental orientations during a wing beat: viewed in the fuselage body fixed reference frame CB0.

experimental flight platform in free flight was visualized. Large deflections occurring during the free flight can be seen in the fuselage fixed reference frame CB0. Fig. 6.7 shows 3 out of 34 example position states of the wings in free flight in the fuselage reference frame CB0.

For orientation, the fuselage and the luff and thrust flap region are indicated in Fig. 6.7. The total flapping range and amplitude is 483 mm, the total maximal deformation of the wing in the fuselage fixed reference frame CB0. The flapping amplitude in the following normalizes experimental wing deformation emanating from the wing fixed reference frames to gain prospective on the magnitudes.

6.1.3 Wing Fixed Reference Frames

Elastic deformations modeled in local wing fixed reference frames are optimally fitted in flight test data E-2 to achieve the small deformations that are compliant with linear constitutive material assumptions of the wing components. Modeling assumptions are checked and experimental deformations emanating from the local reference frames are described. Recall that in the simulation, the large global motions are achieved through the induced motion of wing reference frames (underlying rigid-body motion) and the elastic deformation emanating from the wing fixed reference frames.

The following flight test data analysis leads to the five-body model due to remaining large deflections of the wing body fixed reference frames CW in the thrust flap region area. For simulation results to remain accurate, especially when changing model configuration and for verification with flight test data E-2 results, the five-body approach must ensure the validity of linear geometric and material structural dynamic assumptions on the wing.

For the final five-body flexible multi-body dynamics model (FMBDM), local wing fixed reference frames are fitted optimally in flight test data E-2 to adjust underlying rigid-body motion and resulted in small deformations (<10% of the flapping range) of the wing reference frame. Two reference flapping motions in the flight test data E-2 were identified, which can be thought of as underlying rigid-body motions. The criteria for an optimal fit emanating from wing reference frames were to minimize experimental out of plane deformation.

The local wing fixed reference frames are CW attached to wings under zero deflection PS0. For convenience recall the definition wing fixed reference frame CW (Fig. 6.8), where x is along the axis of the fuselage and y is along the wingspan.

The zw-coordinate describes the deformation of the reference frame flapping around x with FA β.

The reference flapping motion of β was optimized such that the smallest deformations of the wing fixed reference plane over the flapping cycle was achieved when fitted in the flight test data E-2.

Fig. 6.9 shows the computed maximal occurring deformation (for this minimal case) of a wing fixed reference plane. The maximum deflection during the flapping cycle of all marker locations in the wing fixed reference frame was computed. Its surface was plotted versus PS0 reference plane. The

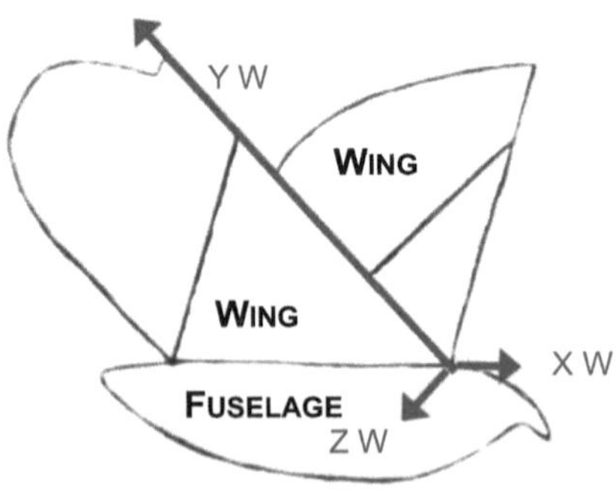

Figure 6.8 Wing fixed reference frame CW.

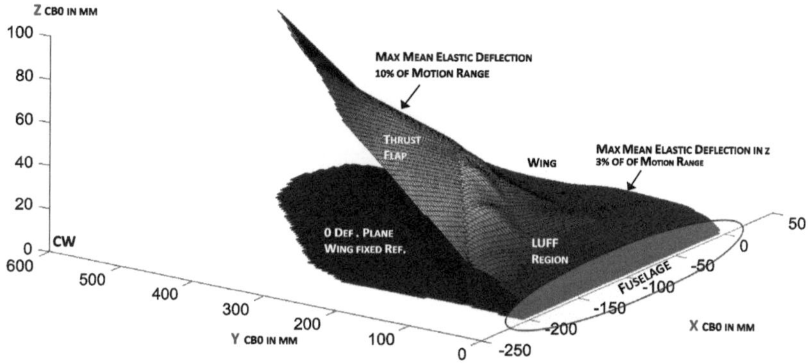

Figure 6.9 Bioinspired ornithopter test platform wing in free flight: maximal elastic deflections of the wing in the wing fixed reference frame during a wing beat; viewed in the fuselage fixed reference frame CB0.

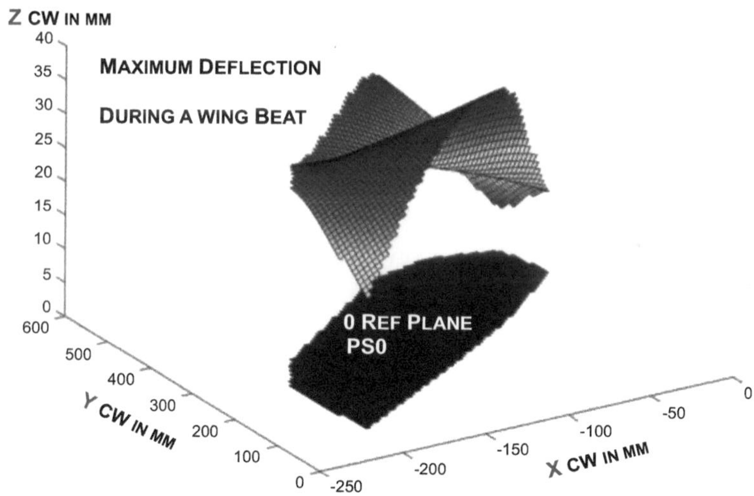

Figure 6.10 Bioinspired ornithopter test platform wing in free flight: maximal occurring elastic deformation in the thrust flap region of the wing in the wing fixed reference frame CW.

maximal elastic deformation occurring during a flapping cycle (mean value over all points) from the wing fixed reference frame in the thrust flap region is 9.54% of the flapping range with a STD of 5.88%. Here the maximal deflection at a point occurring during a flapping cycle is relatively large and 19.59% (Figs. 6.9 and 6.10).

The maximal elastic deflection (mean value over all points) from the wing fixed reference frame in the luff region is relatively small and 3.08%

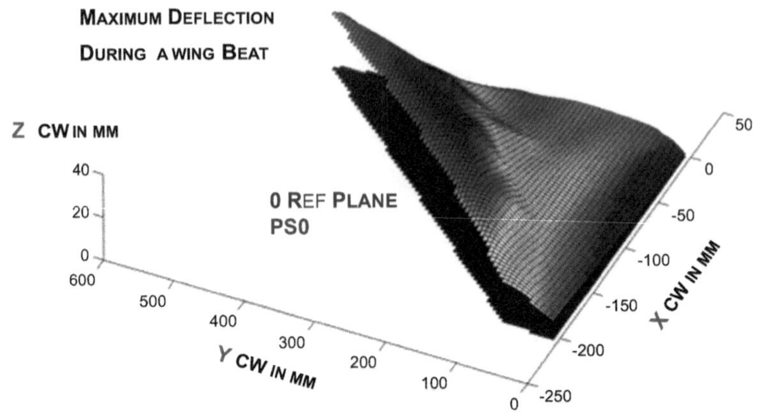

Figure 6.11 Bioinspired ornithopter test platform ML101 wing in free flight: maximal occurring elastic deflections luff region in the wing fixed reference frame CW.

of the flapping range with a STD of 1.96% and shown in Fig. 6.11. The maximal deflection point occurring during a flapping cycle is 7.49% of the flapping amplitude. The maximal deformations are small here and linear elastic material constitutive assumptions are valid.

Owing to the large maximum elastic deflection in the thrust flap region shown in Figs. 6.12 and 6.13 an additional thrust flap region fixed reference plane was required to model the ornithopter system using a linear elastic flexible multi-body dynamics assumptions.

An optimal fit of a thrust flap region reference plane reduced the deformation in the thrust flap region of the wing fixed reference planes significantly. The maximal mean of all tracking markers elastic deflection of its reference frame reduces from 9.54% to 5.21% of the flapping amplitude,

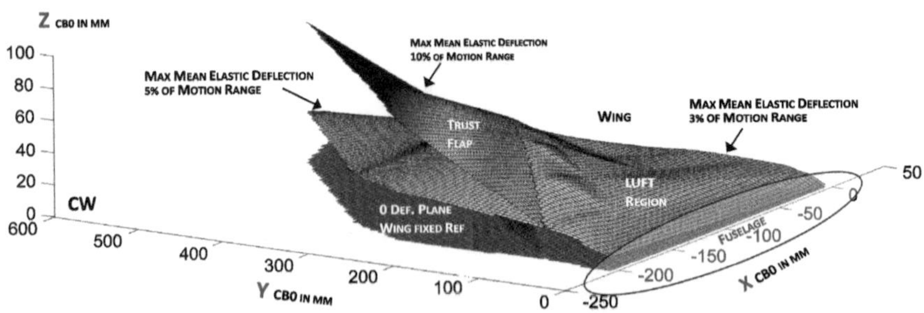

Figure 6.12 Bioinspired ornithopter test platform wing in free flight: maximal occurring elastic deflections of the wing fixed reference frame during a wing beat; viewed in the fuselage reference frame CBO.

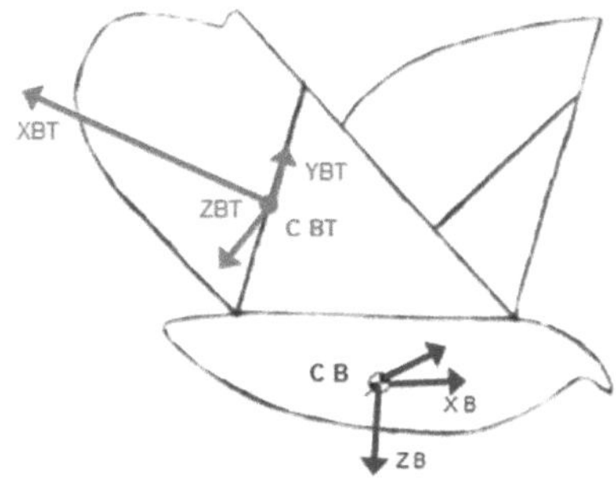

Figure 6.13 Schematic location thrust flap region reference frame CBT.

with an STD of 2.12%. The reductions of mean maximal elastic deformation of the zero reference plane PS0 are shown in Fig. 6.12.

Minimal average deformation emanating from the wing fixed reference frames (computed overall position states during a flapping cycle) was achieved by an additional thrust flap reference motion. The position states of the x-axis of the thrust flap fixed reference frame CBT at the flapping frequency β of 6.06 Hz are shown in Fig. 6.14. For convenience, recall the definition of the thrust flap fixed reference frame (Fig. 6.13). The z-location of a marker point is deformations emanating from the thrust flap fixed reference frame.

xBT in Fig. 6.14 is the x-axis of the thrust flap region fixed reference frame plotted in 34 position states and is the underlying rigid-body reference motion in five-body flexible dynamics model. Shown here is the location of

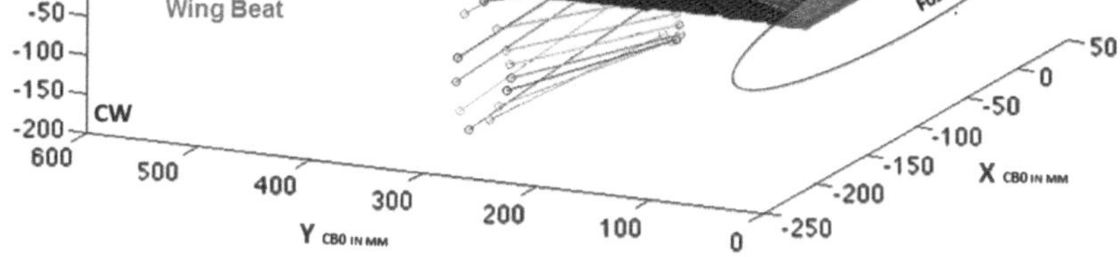

Figure 6.14 Bioinspired ornithopter test platform wing in free flight: X-axis of modeled thrust flap region reference frame in 34 position states viewed in the fuselage fixed reference frame CBO.

the xBT is the optimal fit of orientation and location of the rigid reference motion to archive the smallest deformations of the wing fixed reference plane PS0. Z- and Y-axes of CBT remain attached in orientation and location to the wing fixed reference frame CW.

Reference motions and optimal fit (minimal deformation of the wing fixed reference plane in z) was found by fitting the x-axis of CBT between location marker 28 and marker 30 on the wing. The z-axis and y-axis location of the reference frame CBT in reference frame CW is constant over the flapping cycle and accordingly its PS0 position. The optimal fit of the reference motion (angle zeta ζ) shows the minimal deformation of the thrust flap region reference plane of 5.21% of the flapping range. The tested underlying rigid-body motion combinations were a rigid x-axis fit of CBT between maker locations (42/53, 42/52, 46/53, 46/52, 47/51, 48/51, 49/51, 29/31, 28/31, 27/30, 25/30, 25/26, and 25/9) over the flapping cycle. Therefore, the FA (zeta ζ) position states over a flapping cycle results from flight test data according to Fig. 6.14. The position on the wing of the x-axis (reference frame CBT) is shown in Fig. 6.15, a one DOF motion around the y-axis of CBT.

One, two, and three DOF of the underlying flapping motion was tested for the previously mentioned combinations of marker locations. However, additional DOFs showed no significant improvement in the magnitude of deformation emanating from the reference plane and would unnecessarily complicate the model. The remaining DOF is and θ_{IM1} in FMBDMs generalized coordinate $\xi_{II(R/L)}$ for Bodies II according to preceding nomenclature. Fig. 6.16 shows the thrust flap angle zeta location on the wing. For one DOF zeta rotates around the y-axis of CBT and is θ_{IM1} in the model.

Fig. 6.17 shows the wing FAs of the underlying rigid-body motion of the five-body flexible dynamics model in comparison to inflight (E-2) found

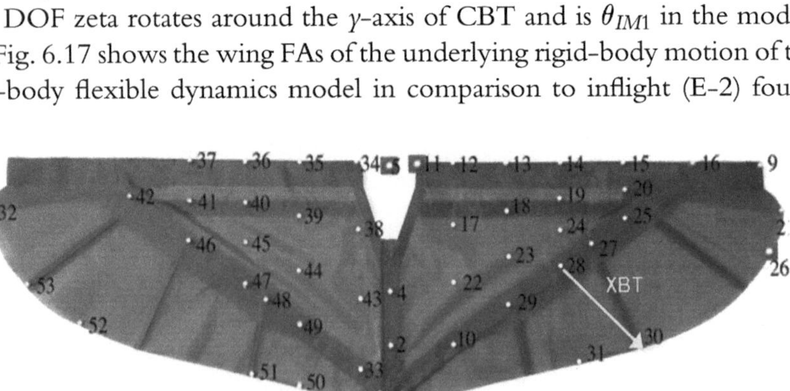

Figure 6.15 Bioinspired ornithopter test platform wing in free-flight marker location: XBT-axis location on experimental test platform resulting in minimal deformation of a reference YBT/ZBT reference plane using a 1 degree of freedom flapping motion.

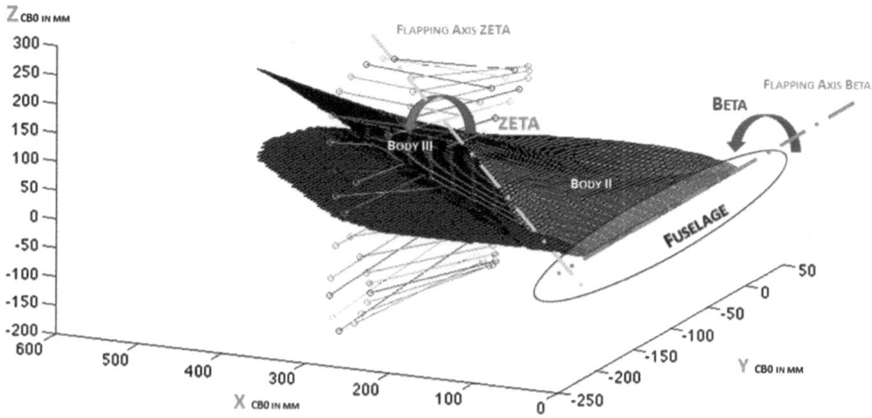

Figure 6.16 Flapping angle zeta and beta location on shown in experimental wing test data (maximal deformation of CW,0 reference plane, and XBT position states).

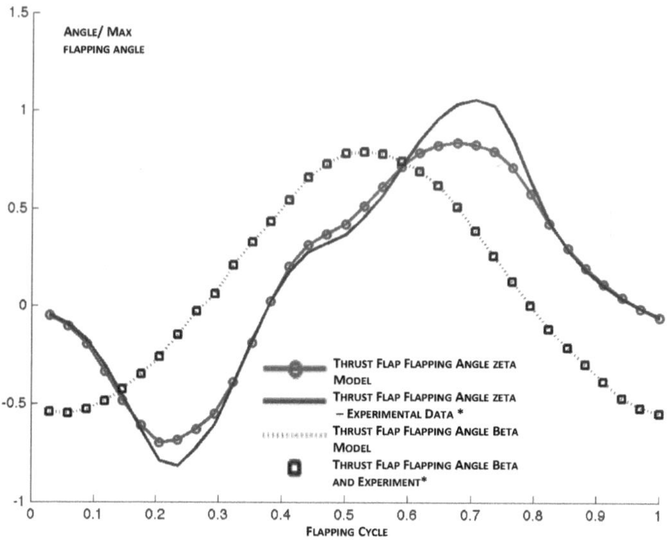

Figure 6.17 Modeled flapping angle (FA) time history based on flight test data versus FA in flight test data E-2.

wing motions. The mean delta zeta is 1.44% of the FA range with an STD of 8.58% of the FA range and the mean delta beta is 0.35% of the FA range with an STD of 5.43% of the FA range. The small difference is due to "wobbling" and small flexibility in the x and y plane in CW and CBT, respectively found

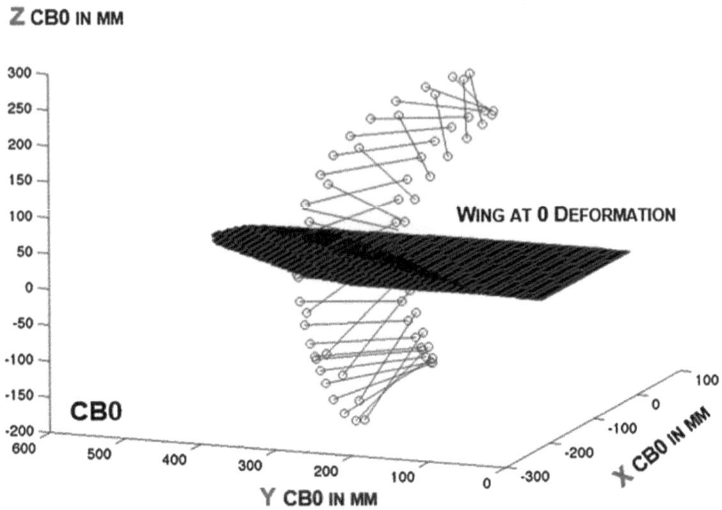

Figure 6.18 Experimental thrust flap reference motion in 34 position states in the fuselage fixed reference plane CB0.

in the flight test data E-2. This x/y plane flexibility was neglected in the underlying rigid-body motion of the model according to Fig. 6.17.

Fig. 6.18 shows thrust flap motion zeta in the experimental data in the body fixed reference frame CB0 at 34 position states over the flapping cycle and had the three rotational DOFs.

The modeled thrust flap reference motion according to Fig. 6.19 has one DOF $= \theta_{IM1}$ and resembles a figure eight motion.

The mean normalized error over 34 position states of the XCB0 coordinates of the thrust flap region x-axis XBT is 2.64% with an STD of 1.51% of the range of motion in z.

The mean normalized error over 34 position states of the YCB0 coordinates on the thrust flap region x-axis XBT in y is 6.07% with a STD of3.83% of the range of motion in z.

The small error shows that one rotational DOF θ_{IM1} is significantly dominant in the working ornithopter platform thrust flap region and additional rotational DOFs can be neglected using an optimal thrust flap region reference frame fit. The dominant motion in the thrust flap region of the working ornithopter flight platform can be described accurately by a figure eight motion that represents the underlying rigid-body motion of the flexible wing in a five-body flexible dynamics model. Recall that the superimposed elastic deformation leads to full deformation found in experimental flight test data.

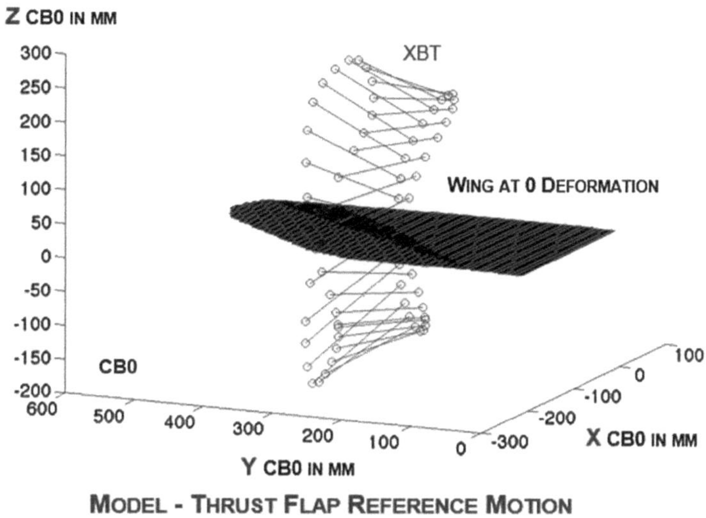

Figure 6.19 Modeled thrust flap reference motion in 34 position states in the fuselage fixed reference plane CB0 versus wing at zero deformation (PS0).

6.1.4 Leading Edge Spar Flexibility

Next, the leading edge spar deformation of the optimal wing fixed reference plane CW is examined. Fig. 6.9 shows the leading edge spar deformation of ML101 during the up and downstrokes and the underlying rigid-body motion in the model that is the *x*-axis of CW according to Fig. 6.8. *X*-axis of CW, the underlying rigid-body motion of the wing and experimental wing deformation are shown in 34 position states during up and downstroke. Experimental wing deformations that are a polynomial fit of the deformation of leading edge markers 5/11/12/13/14/15/16/9 shown in Fig. 6.14. Fig. 6.20 shows the modeled and experimental deformation in the fuselage body fixed reference frame CB0 normalized over the wing half span. The modeled reference motion describes the global large deformation only, which is motion of reference frame CW in the FMBDM. Experimental kinematics shows the magnitude of the elastic deformation and underlying global large deformations. A comparison was made to check the validity of the small deformation assumption of the global reference plane in the model and estimate the leading edge spar flexibility.

In Fig. 6.20, it can be observed that the experimental deformations of the reference plane (model underlying wing reference frame) are small. To examine this closer, switch to the wing fixed reference frame CW.

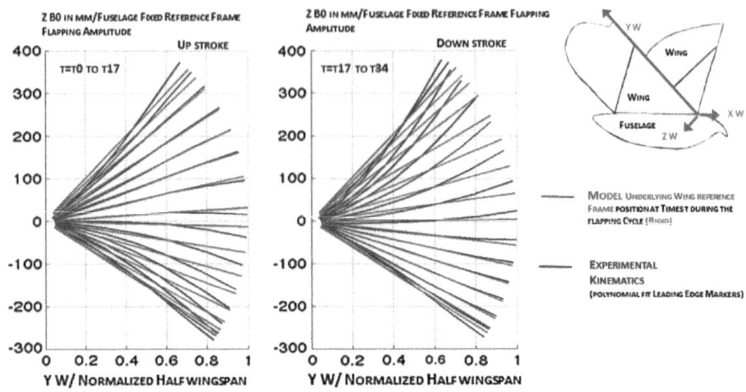

Figure 6.20 Experimental elastic deflections in the fuselage fixed reference frame CB0 during a wing beat.

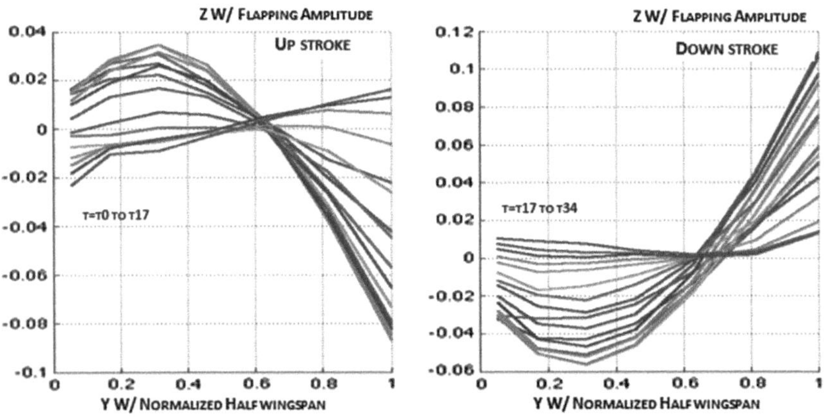

Figure 6.21 Experimental elastic deformation in the wing fixed reference frame CW during a wing beat.

Fig. 6.21 shows the experimental deformation of the leading edge spar in the wing fixed reference plane during a flapping cycle. The flapping cycle is shown in 34 position states during the up and downstrokes and the normalized z-coordinate of the experimental data is shown along the wing half span during those position states. To get a reference magnitude, the z-coordinates were normalized by the flapping amplitude. The maximal deformation due to flexion of the leading edge spar is 11% of the flapping amplitude at position state $t = 34$.

This affirms that small deformations of the leading edge spar in free flight of the wing fixed reference frame CW are observed during the flapping cycle, and linear material constitutive assumptions are valid for the wing model leading edge spar.

In summary, due to the optimal fit of an additional underlying rigid-body motion, small deformations of the wing fixed reference frames were present and linear elastic structural dynamics modeling assumptions for the wing lead to valid results for integration in the FMBDM. Figure eight motion in the thrust flap region was dominant and inherent in the ornithopter test platform ML101 during free flight.

6.2 MODEL RESULTS

The model input is the wing motion profile provided by the flapping motion function on the wing root, which results in dependent thrust flap dynamics according to bioinspired experimental wing kinematics. Further input includes wing stiffness and wing geometry. See Fig. 6.22 for model image.

The model output for each body in the system is the kinetic energy magnitude, translational kinetic energy magnitude, angular kinetic energy magnitude, translational momentum (magnitude, x,y,z), angular momentum about CM, potential energy, center of mass (CM) position, CM velocity, CM acceleration, CM angular velocity, and CM angular acceleration (magnitude, x,y,z) for each body in the system. Such bodies are the fuselage, the two wing thrust flap regions (L/R), and the two wing Luff regions (L/R). For the flexible wing bodies of the system, the mode shapes and strain energy are included in the output.

The model output also included each body attachment point in the system such as the origin of the wing root reference frames CBW and the

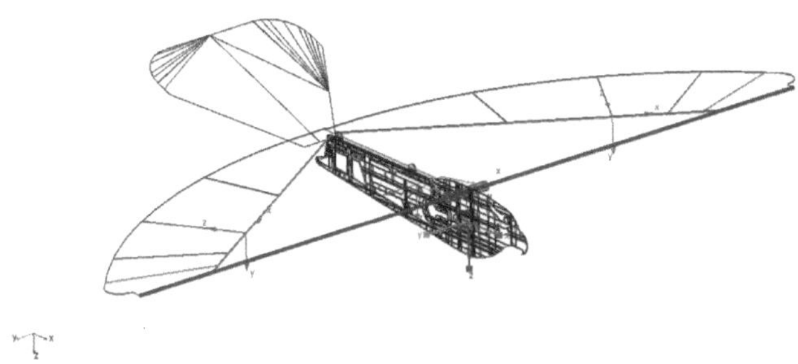

Figure 6.22 Model image SimXpert—ML101 five-body flexible multi-body dynamics model (fuselage reference frame CB is highlighted).

origin of the thrust flap region reference frame CBT. The characteristics obtained were the element force (magnitude, x,y,z), element torque (magnitude x,y,z), translational displacement, projection angles (Ax,Ay,Az), the translational velocity (magnitude, x,y,z), the translational acceleration (magnitude, x,y,z), the angular velocity (magnitude, x,$y = 0$,z), and angular acceleration (magnitude, x,$y = 0$,z).

The model outputs for each chosen marker location in the model are translational displacement, projection angles (Ax,Ay,Az), translational velocity (magnitude, x,y,z), translational acceleration (magnitude, x,y,z), angular velocity (magnitude, x,$y = 0$,z), and angular acceleration (magnitude, x,$y = 0$,z). These output characteristics were chosen and obtained for various input characteristics. Input characteristics according to experimental flight platform wing tip kinematics and integrated loads resulting from model simulations were compared to experimental data (E-1, E-2) in the remainder of this chapter. Loads are obtained in the vehicle dynamics model at the fuselage center of mass CB, which is the highlighted coordinate system in Fig. 6.22.

Individual model verification of disciplines comprising and resulting in the final multi-disciplinary model is also presented. These are the final verifications resulting from model iterations of the independent models of the coupled problem. The aerodynamics model C verification results and FMBDM results without aerodynamic loads are presented in comparison to experimental data in E1-I and E3, respectively. Structural dynamics in FMBDM was evaluated earlier in Section 6.1.

6.3 CONSTRAINT MODEL VERIFICATION

6.3.1 Integrated Forces of the Avian Flight Robotic System

Observe the integrated forces of aero-model A and inertia forces due to the wings in comparison to experimental data E1 (Fig. 6.23) below. That is the constraint configuration of model and experiment shown in Fig. 6.23.

The mean absolute error comparing the VPF E1 to aero-model A is 2.9611N with a mean error of 1.01N/STD 3.68N. The magnitudes of inertia forces due to the wing VFI are comparatively small and the main difference in integrated force results in both (experimental and simulation of constrained vehicles) stems from the aerodynamic modeling assumptions used. Aero-model A uses three blade elements for the proof of concept study of modeling methodology.

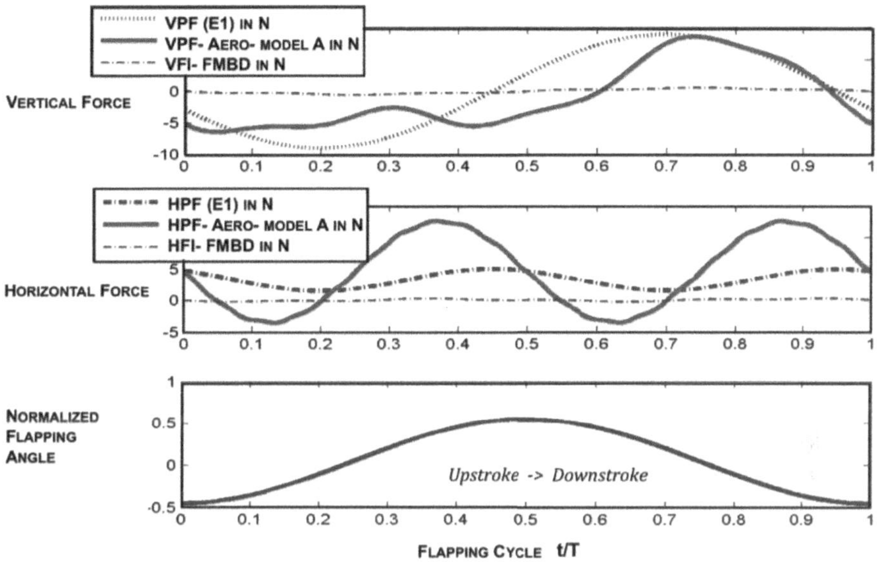

Figure 6.23 Verification bench test—model 3 A—scale ML101 (6.17 Hz). *FMBD*, flexible multi-body dynamics; *HFI*, horizontal force inertia; *HPF*, horizontal propulsive force; *VPF*, vertical propulsive force.

The mean absolute error in horizontal direction indicates an overprediction due to the circulatory force assumption used (according to Harmon). The mean absolute error in the horizontal direction is 4.59N with a mean error of 1.25N and an STD of 5.03N. This proved the methodology of the experimental kinematics–based aero-model A in the FMBDM. Using three blade elements, tracking behavior to experimental data using quasi-steady aerodynamic assumptions worked considerably well. Originally, the error was also thought to stem from neglected inertia forces in an aerodynamics model. Inertia forces have been found to be considerably small and the main contributions of error stem from aerodynamic assumptions. Model A was used in combination with an FMBDM with only one wing reference frame CW. Three body representation (one wing body of each wing) resulted in the main error in tracking wing kinematics (such as wing tip kinematics) and discussed earlier in Section 6.1.3.

6.3.2 Wing Tip Kinematics

In the previous section, the experimental deformation of the wing was compared with the large global deformation resulting from model simulation, the underlying wing fixed reference plane position. In the following, experimental deformation is compared with the total deformation in the model that is the result of the underlying reference frame motion and the superimposed elastic deformation and model performance is discussed. One chooses the wing tip location in 34 position states to compare modeled wing kinematics to experimental wing kinematics. The wing tip position state is determined by the combination of motion of both wing fixed reference frames and the elastic deformation in the model, lies of axis of any reference frames used and is inherently one of the most extreme deformations on the wing making it a good geometric location for a kinematic verification point on the wing. The wing tip position is the marker two location indicated in Fig. 6.15. See Fig. 6.24 for wing tip location in one experimental position state during free flight of the wing during downstroke.

The blue solid line in Fig. 6.25 shows the experimentally measured wing tip path (E-2) using 34 position states during one flapping cycle. The red interrupted line shows 34 position states of the five-body flexible dynamics model where aerodynamic loads result from final Aerodynamic Model C. The wing tip path in free flight and the simulation results in free flight are shown in fuselage body fixed reference frame on the wing root CB0.

The magnitude error of the position state in the y/z plane (back view) is 3.04% with an STD of 1.42%. The magnitude difference in the position state in the z/x plane (side view) is 5.43% with an STD of 4.51% over all position states. The data are normalized by its maximal deformation occurring in x, y, and z, respectively. The maximal elastic deformation of the wing tip location during the wing beat of the thrust flap region fixed reference plane is 7.49% of the flapping range. The magnitude of the error comes from the error in elastic deformation of the wing of the underlying wing reference

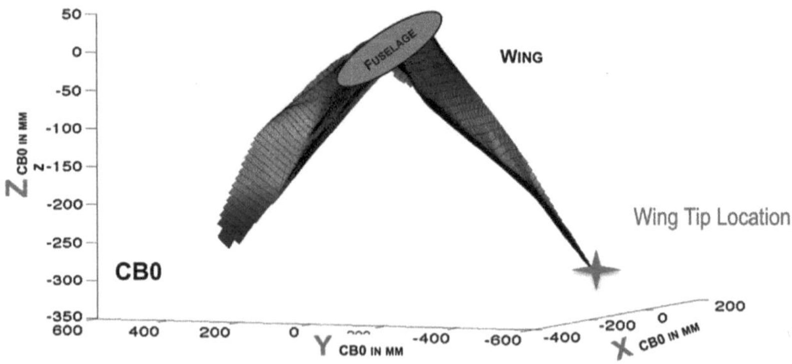

Figure 6.24 Wing tip location—viewed at one experimental position state (mid-downstroke).

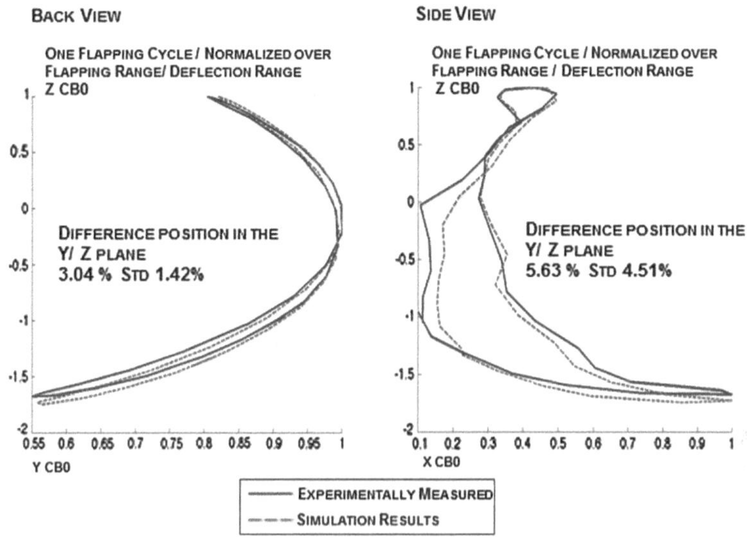

Figure 6.25 Results wing kinematics model 5 experiment: flight test: (left) back view and (right) side view.

frames. Owing to the kinematics and location of wing fixed reference frames, elastic deformations are mainly directed in ZB0; therefore, only the z/y and z/x plane CB0 is used for verification purposes.

These elastic deformations are the result of geometric and stiffness properties of the wing and aerodynamic load magnitude and distribution.

6.4 UNCONSTRAINT MODEL VERIFICATION

6.4.1 Integrated Forces

The following addresses the model verification in free-flight condition. The fuselage is here unconstrained. Vertical and horizontal forces observed at the fuselage center of mass at the origin of CB were observed in free-flight test data (E-2) and in the unconstrained five-body vehicle dynamics model using aero-model C. Model verification results are shown in Fig. 6.26. E-2 data used were previously discussed in Section 3.2.3 (Fig. 3.32) in Chapter 3.

The mean error in the vertical direction is 2.34N/STD 11.70N (mean absolute error is 9.505N/STD 5.5518N). Observed forces correlated well. A slight under prediction in VF was observed and assumed to stem from aerodynamic model assumptions. The mean error in the horizontal direction is 3.758N/STD 2.6N (mean absolute error is 2.08N/STD 4.37N). The

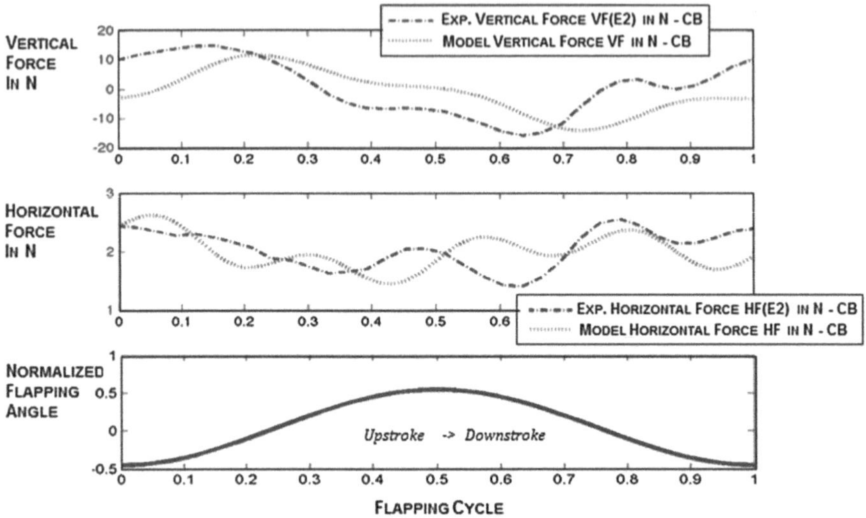

Figure 6.26 Propulsive forces acting on the fuselage center of mass—obtained from experiment (E-2) and five-body vehicle dynamics model using aerodynamic model C.

vehicle was flying with a mean flight velocity, and forces observed were due to acceleration in flight direction during the flapping cycle. Vertical forces were observed at the fuselage center of mass due to the present heave motion in the flight platform (ML101) during free flight. Forces on the fuselage center of mass CB gave a good estimate about model performance when compared with free-flight test data.

6.4.2 Wing Inertia Forces

The total integrated forces acting on the origin of fuselage fixed reference CB due to the inertia of the wing are discussed next. Two cases are compared: Case 1: with the thrust flap motion zeta turned off (a wing with one flexible body) and Case 2: with the thrust flap motion turned on and two wing bodies according to final five-body vehicle dynamics model. The results were computed at a flapping frequency of 6.06 Hz and shown in Fig. 6.27.

Inertia forces exerted on the fuselage center of mass CB in free flight are smoother using a two-body wing simulation, which corresponds to the two section luff and thrust flap region wing versus a wing using only one body. Inertia forces due to the wing are of small magnitude compared with aerodynamic forces. Besides the aerodynamic advantages, division is advantageous considering inertial forces exerted on the vehicle fuselage only.

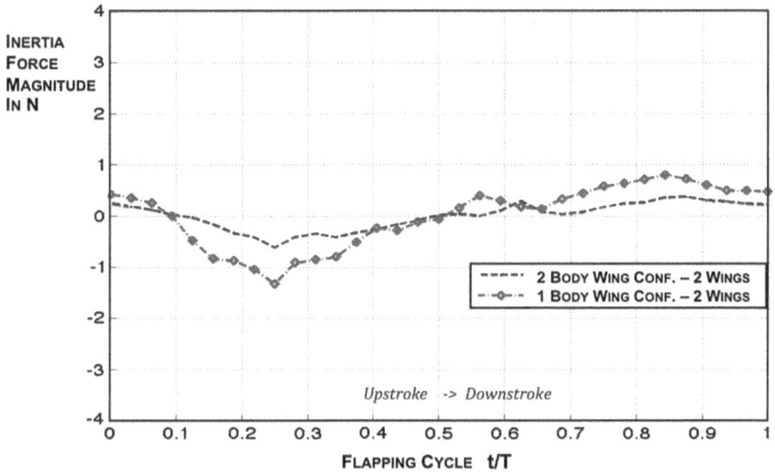

Figure 6.27 Simulated inertia forces on fuselage center of mass CB in free flight (inertial vertical propulsive force)—upstroke–downstroke transition.

Using the one wing body model, the mean value of inertia forces is −0.0054N (STD 0.26N) with a mean absolute value of 0.21N (STD 0.13N). For the two-body flexible wing model, the mean of the inertial force is −0.0105N (STD 0.06066N), and the mean absolute value is 0.5108N (STD 0.13N). In summary, the inertial forces are smoother using the two-body flexible wing representation according the five-body FMBDM.

The magnitude of forces resulting from the five-body dynamics model was compared with the experimental data (E3) (Fig. 6.28). It was observed that the mean absolute value of the simulation lies well within the magnitude of measured inertial vertical propulsive forces. A simulation ran at 4.7 Hz flapping frequency of (FA-beta) to obtain a comparison operation point for experimental data set E3, and at the baseline flapping frequency used throughout this work, 6.06 Hz. Owing to the large measurement error in E3 for this small magnitude of forces, the verification mainly served to do a magnitude check of the occurring inertia forces in the vertical direction.

A further validation of inertial horizontal propulsive force and inertial pitching moment adds no additional value to the credibility of the model, but one can observe that measured and simulated forces (averaged over the flapping cycle) are comparable and of same magnitude. Therefore, the five-body dynamics model predicts inertia forces accurately. This depended on the masses, mass distribution, orientation, and acceleration of the model components. Orientation and accelerations are dependent on the flapping

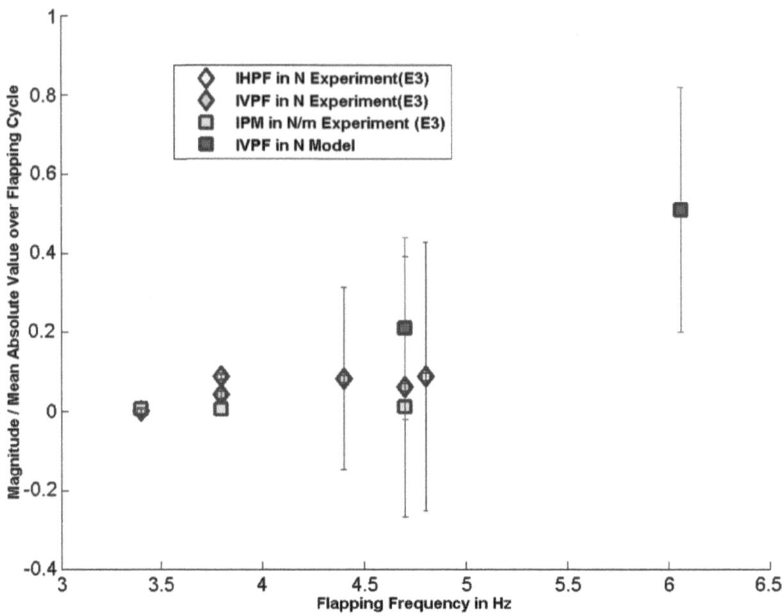

Figure 6.28 Results inertia forces (E3) with error bars versus simulation results with STD, (A) inertial horizontal propulsive force (IHPF), (B) inertial vertical propulsive force (IVPF), (C) inertial pitching moment (IPM) versus magnitude.

frequency (beta), dependent thrust flap angle (zeta), and wing reference frames, which are according to flight physics of the test platform. Since the geometry and mass properties of the model components are according to test platform, the mass and mass distributions also lead to accurate results of the FMBDM. The elastic deformation and therefore the inertia change due to elastic deformation during the flapping cycle without the presence of aerodynamic loads are small. Elastic deformations are mainly caused by the aerodynamic forces which are 45 times larger than occurring inertia forces for ML101 (6.06 Hz).

6.4.3 Inertia Invariants

Recall that the floating frame of reference formulation resulted in a highly nonlinear mass matrix because of the inertia coupling between the reference motion and the elastic deformation [1]. According to discussion in Section 4.1, three cases were implemented to quantify the effect of inertia coupling due to the elastic deformation on the wing. EOMs for the five-body dynamics system of the ornithopter were solved for the cases and compared with the experiment. Here Case (A) was the rigid-body motion only.

Case (B) included no inertial coupling between rigid and flexible body EOMs and Case (C) where the EOMs were fully coupled.

Case (B) lead to a 20.56% (STD 8.64%) difference in the integrated load results compared with Case (A). This was computed in CB0 for the flapping frequency of angle β of 6.06 Hz and recorded over the flapping cycle. Case (B) correlated better to experimental data but was inconclusive and the differences were within the measurement error for the inertia forces due to the wing. Case (B) is thought to be an improvement over Case (A) due to its higher fidelity. However, this showed that inclusion of wing flexibility using flexible wing structure is not highly significant for the inertial force generation. Wing flexibility is mainly significant for aeroelastic effects in the five-body vehicle dynamics model. The effect of the inclusion of a second wing body in the wing model was more significant and resulted in a 56% (STD 30.81%) difference in inertial forces due to the wings. This was measured in CB0 of the five-body rigid vehicle dynamics system versus CB0 in the three-body rigid vehicle dynamics system (for FA β of 6.06 Hz over the flapping cycle). In the three-body system the thrust flap regions was rigidly attached to the luff region of the wing.

Including the inertia cross terms Eqs. (82) and (89) in Section 4.6.5 for the wing bodies according to Case (C) lead to a 10.56% (STD 5.64%) difference for the inertial loads due to the wings compared with Case (B) on CB0 (for FA β of 6.06 Hz over the flapping cycle). The inclusion of inertia cross terms according to Case (C) became more significant with higher flapping frequency, increased aspect ratios of the wing, increased mass, and increased deformation of the wing of their reference frames. They were included in the five-body vehicle dynamics model to facilitate such a change in configuration.

6.4.4 Aerodynamic Forces

Next, the aerodynamic forces obtained from final aero-model C were compared with aerodynamic forces obtained from system ID through experiment (E1-I). This was the decoupled force verification for aerodynamic forces in free-flight conditions according to assumptions and constraints of aerodynamic model C.

To provide comparability to source data, the data (E1-I 3.2.2) was plotted over the wing stroke and normalized over the wing stroke due to the scaling invariance of test platform MSK004 (E1-I) and aero-model C tailored to MSK001 (Fig. 6.29).

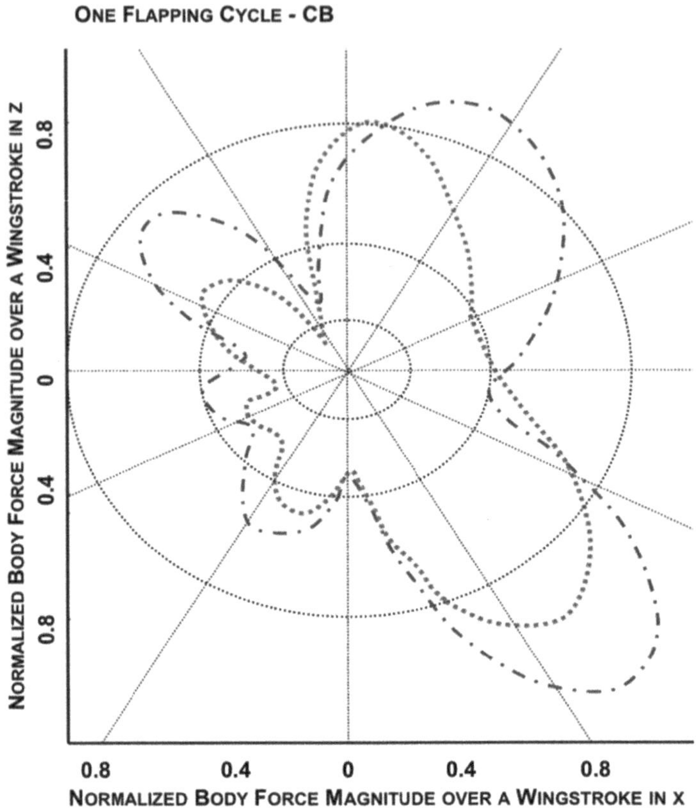

Figure 6.29 Results: body force magnitude system ID model (MSK004 E1-I) [4]—body force magnitude—aero-model C.

The normalized error (mean over wing stroke/normalized by max force sys ID model) is 10.71% of the max system ID model force with an STD of 9.87%. Good tracking behavior of aerodynamic model C (based on blade element theory) was observed compared to the model obtained by system identification. It was assumed that difference in magnitudemainly stems from differences in wing scale (MSK004 to ML101).

Note that in workflow Step 2, validation and model iteration (Fig. 11 in Section 2.2), the aerodynamic model output was time-variant and spatially distributed (TVSD) force. Integrated forces and moments were correlated to integrated forces and moment data obtained by Grauer through systems identification of free-flight test data using the secondary test platform Odyssey MSK004 [9]. Integrated horizontal and vertical propulsive forces and moments were obtained at the fuselage center of mass through system identification and aero-model C.

Within its limitations, the data set by Grauer provided the appropriate and comparable data set for decoupled aerodynamic forces in free flight. Integrated forces and moments in clamped conditions were also obtained for the primary and secondary test platform by Harmon, due to numerous factors not suitable for the correlation of the aerodynamic model. For example, forces and moments contain inertial effects and circulatory effects change the aerodynamics in clamped conditions as observed in these tests [2].

Correlation of the aerodynamic model output can provide a good estimate of the quality of the aerodynamic model assumptions. However, an infinite number of solutions of TVSD forces on the wing can lead to the same horizontal propulsive and vertical lift force. A scaling difference between the test platforms used for the wing kinematic input (primary test platform) and the system identification data (secondary test platform) also influenced the aerodynamic. Scaling is still a significant challenge in flapping wing aerodynamics and its influence is not entirely understood [3], leading to a magnitude of differences between integrated forces of the model and system identification model data by Grauer. Owing to the infinite number of solutions and a scaling difference, an optimization algorithm for aerodynamic model improvement was not advantageous, and model iterations were performed using an educated model tuning. Aerodynamic model assumption changes were made according to known influence factors significant to flapping wing flight and observed through open literature or empirical knowledge. Model iteration changes entailed changes in lift distribution on a blade element, dynamic stall cut off angles, and spanwise lift distribution. The data set by Grauer provided the only available comparison metric in free literature to give a good estimate of the quality of aerodynamic modeling assumption as a stand-alone in the model development process.

REFERENCES

[1] A.A. Shabana, Flexible multibody dynamics: review of past and recent developments, Multibody System Dynamics 1 (1997) 189–222.

[2] R.L. Harmon, J.E. Hubbard, Aerodynamic Modeling of a Flapping Membrane Wing Using Motion Tracking Experiments (Master's thesis), 2008.

[3] W. Shyy, Y. Lian, J. Tang, D. Viieru, H. Liu, Aerodynamics of Low Reynolds Number Flyers, Cambridge University Press, 2008.

[4] J.A. Grauer, J.E. Hubbard Jr., Flight Dynamics and System Identification for Modern Feedback Control: Avian-Inspired Robots, Woodhead Publishing in Mechanical Engineering, 2013. ISBN-13: 978-0857094667.

CHAPTER SEVEN

Concluding Remarks About Modern Modeling Methodology Implementation and Flight Physics of Avian Scale Flight Robotics Systems

7.1 SUMMARY OF MODERN MODELING METHODOLOGY DEVELOPMENT AND IMPLEMENTATION

A novel modeling methodology and a five-body flexible dynamics model considering aeroelasticity were described in the body of this work. Structural dynamics and linear elastic modeling assumptions were verified and compared with in-flight wing kinematics data measured for this purpose, as well as an array of arrival experimental data. Model output correlated to experimental test data, thus the model is available for utilization in the multi-disciplinary design optimization of a flapping wing ornithopter. This work contributed a nonlinear vehicle dynamics model of an ornithopter using a linear elastic multi-body dynamics representation of an ornithopter with a quasi-steady aerodynamics model (initialized by free-flight kinematics) to consider aeroelasticity of the wings. Linearizing nonlinear structural dynamics of the wings due to large deformations during free flight was also explored by investigating free-flight kinematics. A unique modeling methodology was developed to facilitate the model.

Chapter 1 described the background for the areas of interest in this work. This included background in flapping wing, flapping wing aerial vehicles, low Reynolds number, unsteady aerodynamics and aerodynamic models for flapping wing, flexible multi-body dynamics, vehicle dynamics, and models of ornithopter. Chapter 2 described the modeling methodology developed in this work. Chapter 3 described the experimental test platform, archival data, and experiments performed within the assumptions and constraints of the model. Chapter 4 described the flexible multi-body dynamics model developed, and Chapter 5 described the aerodynamics

Modern Flexible Multi-Body Dynamics Modeling Methodology for Flapping Wing Vehicles
ISBN: 978-0-12-814136-6
http://dx.doi.org/10.1016/B978-0-12-814136-6.00007-X

model developed. In Chapters 4 and 5, model integration within the aforementioned modeling methodology was also outlined. Chapter 6 showed and discussed results of verification of modeling assumptions and simulation results.

7.2 SCOPE AND CONTRIBUTIONS RESULTING FROM MODERN MODELING METHODOLOGY IMPLEMENTATION DESCRIBED IN THIS BOOK

The main contribution of the work described here is the development of an experimentally verified aeroelastic flexible multi-body vehicle dynamics model of a bioinspired ornithopter using a novel modeling methodology. The multi-disciplinary model results of models in individual disciplines and their iterations that involved aerodynamics, structural dynamics, and multi-body dynamics. The problem of ornithopter comprised of unsteady aerodynamics in a low Reynolds number flow field, flapping wing aerodynamics, fluid—structure interaction (FSI), and nonlinearities in both structural and multi-body dynamics. Experimental data obtained and analyzed for the understanding of the flight physics and dynamics of avian scale flapping wing flight aided the model development.

The scope and contributions of this work can be subdivided into the following areas of interest: contributions to flexible multi-body dynamics modeling of an ornithopter, flexible multi-body dynamics modeling, aerodynamics modeling of flapping wing flight, aeroelasticity in flexible multibody dynamics modeling vehicle dynamics modeling, multi-disciplinary modeling methodology, and understanding of flight physics and dynamics of flapping wing flight. Novelties within this body of work are highlighted subsequently.

7.2.1 The Vehicle Dynamics Model of Ornithopter

In summary, a five-body flexible dynamic model of bioinspired ornithopter resulted and is described in this book. The details of the model are as follows:

- Vehicle dynamics model considered wing flexibility, multi-bodies, and FSI in the wings of a bioinspired ornithopter.
- A floating reference frame formulation was used where two sets of coordinates described the configuration of a deformable wing body.
- Elastic deformations were described in local wing fixed reference frames. The local wing fixed reference frames were attached to wings under zero deformation.

- Wing fixed reference frames were actuated to achieve large global motions (large flapping motion) and oscillating body dynamics.
- Structural deformations were obtained by a linear combination of mode shapes. The implementation of flexible body constraints and modes used the Craig–Bampton method.
- Inertial coupling through elastic deformation of the wing was considered using cross-terms in the inertia matrix. This allowed simulation accuracy in changing the ornithopter configurations (mainly if wing mass or flexibility was significantly increased).
- The kinematics obtained in experimental free flight initialized the model.
- The model inputs were experimental wing kinematics (initialization of aerodyamic loads), wing stiffness, wing geometry, material properties, and the flapping motion at the wing root (that is the flapping angle beta). The model outputs were deformations of the vehicle, kinematics, dynamics, and energy. These in addition to modes, stresses and strains on the wings can be obtained for each component of the system or the system as a whole.
- The structural dynamics, multi-body dynamics, and aerodynamic modeling assumptions have been experimentally verified.
- The multi-disciplinary model was implemented using Matlab (aerodynamics), Patran/Nastran (structural dynamics), and Adams (multi-body dynamics) in a SimXpert framework, and integrated simulation was archived.

7.2.2 The Aerodyamics Model and Aeroelasticity

In summary, three analytical aerodynamic model iterations (aerodynamic model A, aerodynamic model B, and aerodynamic model C) using experimental wing kinematic inputs were developed. The final aerodynamic model C was used for FSI in the five-body flexible dynamics model. Final Aerodynamic Model C was an efficient and practical application for the aerodynamic and aeroelastic analyses of flapping wings. The scope of Aerodynamic Model C is further discussed next.

The model was formulated in terms of genereralized coordinates of the five-body flexible vehicle dynamics system.

The model considered cord- and spanwise flexibility in the wing, due to camber and orientation in each blade element (BE), and fore-and-aft approach in wing-cord direction.

The model used a BE refinement of 13 blade elements per wing, 4 in the thrust flap region, and 9 in the luff region of the wing. For the model development, experimental wing kinematics over a wing stroke served as input variables. In the final model, kinematics obtained from the five-body flexible dynamics model served as input variables for aerodynamic load generation. The first set of loads was obtained through experimental kinematics.

The quasi-steady and analytical aerodynamic model using BE theory accounted for noncirculatory forces, circulatory forces, leading edge suction, the cord-wise force due to the camber, cord-wise friction drag, and dynamic stall.

7.2.3 The Modeling Methodology

Novel contributions and scope of this work described in this book in modeling methodology are summarized according to the following:

- The individual models were developed in close conjunction with flight-based data, due to the limitation in state-of-the art understanding of flapping wing flight. Individual models were correlated against experimental data and integrated into a comprehensive model.
- The wing kinematics, obtained from experimental data served as input variables for model development, and initialization for fully integrated flexible multi-body dynamics model considering aeroelasticity. The resulting multi-body model was coupled to archive the ability to utilize the model for a changed vehicle configuration. This model was initialized by experimental kinematics.
- The novel model initialization with empirical flight test data lead to accuracy throughout the simulation and fast correlation of a new Flexible Multi- Body Dynamics (FMBD) vehicle configuration when implemented (wing motion profile, wing stiffness, and wing geometry was changed).
- The formulation of the aerodynamic model in terms of the coordinates of the five-body dynamics system was done to fully integrate the model and was novel in the modeling methodology.
- The efficient fluid structure interaction procedure, used superposition of sign waves for the model reduction was novel. The aerodynamic model output was formulated as continuous function which was the exchange variable to the structure. This improved FSI.

7.2.4 An Understanding of Flight Physics and Dynamics

A summary of novel insight into flight physics and dynamics of efficient bioinspired ornithopter is as follows:

- The quantification of flight efficient wing motion profiles through flight test data, that is flapping angle beta kinematics and associated flapping angle zeta kinematics were a new contribution to the field of flapping wing flight.

- The identification of flight efficient wing flexibility through flight test data was novel and aided in the understanding of flapping wing flight physics and dynamics (e.g., flight efficient flexibility of up to 12% of half-span emanating from the wing fixed reference frames during a flapping cycle).
- The analysis of in-flight test data revealed inherent figure eight motion in thrust flap region and identified efficient wing kinematics in avian scale flapping wing flight.

7.3 SUMMARY OF NOVEL CONTRIBUTIONS RESULTING FROM MODELING METHODOLOGY

Novel contributions in this work are highlighted in the following. Original contributions of experimentally verified and integrated multi-disciplinary model of an ornithopter described in the body of this book were as follows:

1. The five-body flexible multi-body dynamics model of an ornithopter itself.
2. The five-body vehicle dynamics model of ornithopter considering FSI.
3. The flexibility consideration in the wing, here the accurate inclusion of wing flexibility in the multi-body dynamics of an ornithopter was novel as well as that the model accounted for cord and spanwise wing flexibility in both aerodynamics and structural dynamics.
4. The model accounted for dynamic elastic inertial coupling.
5. The aerodynamic model was formulated in terms of generalized coordinates of the five-body flexible dynamics system.
6. The model accounted efficiently for FSI through novel model integration and methodology.
7. The model framework was experimentally verified utilizing free-flight test data of ornithopter.
8. The model development was in close conjunction with flight physics and dynamics through empirical observations.
9. The aerodynamic model used in-flight kinematics as input variables.
10. The quasi-steady BE aerodynamic model provided time-variant loads on the wing as a continuous function. These were spatially distributed on the wing. (Mentioned is a novel aerodynamic model reduction technique and results in efficient FSI.)

11. The novel experimental wing kinematics of ornithopter in free flight (E-2), the contribution of this work was the design of experiment for verification of flexible multi-body dynamics model.
12. The verification of individual models comprising the coupled problem of avian scale ornithopter using experimental data, and integration of tools into a fully integrated efficient simulation initialized by free-flight kinematics of flapping wing was novel in modeling methodology.

7.4 SUMMARY OF CONCLUSIONS ABOUT THE MODERN MODELING METHODOLOGY

It was concluded that the linear elastic multi-body dynamics model was appropriate for nonlinear elastic vehicle problem with inherent large deformations due to utilization of appropriate underlying rigid-body reference motions. Fully integrated model utilizing aerodynamic model C led to accurate multi-body vehicle dynamic predictions. The following summarizes the conclusions of this work:

1. The model flexibility assumptions were in tune with flexibility and wing deformation kinematics found in working ornithopter platforms in free flight.
2. The cord- and spanwise flexibility in the wing was considerable. Camber forming in aerodynamic model had to be done over two regions in cordwise direction. Only out of plane deformations in the z-direction off the wing fixed reference frames were considerable but remain small and under 10% of global deformation amplitude. The x/y flexibility in the wing fixed reference planes was negligible.
3. The linear elastic multi-body dynamics model used the Craig—Brampton Method for the integration of the flexible bodies in the multi-body system and lead to accurate predictions for the problem.
4. The inertial coupling through elastic deformation of the wing needed to be considered to allow simulation accuracy in changed ornithopter configurations.
5. A model using in-flight wing kinematics for initialization of aerodynamic loads lead to flight physics based results and fast correlation of flexible multi-body dynamics model considering FSI.
6. Five-body underlying rigid-body motions lead to accurate predictions of ornithopter system using linear elastic multi-body dynamics and lead to a considerable model reduction compared with a nonlinear elastic multi-body model.

7. Two wing fixed coordinate systems resulted in small deformations of the reference frames and therefore diagonalize the stiffness matrix in the equation of motion of the five-body flexible multi-body dynamics system.

7.5 SUMMARY OF RECOMMENDATIONS FOR MODELING OF AVIAN SCALE FLAPPING WING FLYERS

The following recommendations were subdivided into modeling recommendations, verification recommendations, and recommendations investigating flight physics and dynamics.

Model recommended improvements included a time-variant aerodynamic force due to the tail in the flexible multi-body dynamics model of the ornithopter through the coupling of an aerodynamic model. This allows the tail to be used as a control surface in the vehicle dynamics model (e.g., reduce heave and pitch motions).

Further inclusion of verification of the model using its flight trajectory versus experiment (E2) flight trajectory is recommended using translational and rotational position states, velocity, and acceleration of fuselage fixed reference frame CB.

At the expense of computational cost, it is recommended to develop a computational fluid dynamics (CFD) model utilizing in-flight kinematics to refine aerodynamic load distribution and aerodynamic load magnitude on the flexible wings in the model.

Furthermore, it is recommended to utilize the model for flight physics and dynamics studies to study the variation of the wing motion profiles (that is the flapping angle beta) and cord- and spanwise adjustments of wing flexibility (wing stiffness in the model), therefore, improving performance of avian scale ornithopter.

Observing biological flyers and analyzing bioinspired test platform flight test data, it is believed that underlying figure eight motion in the wing plays an important role in unsteady aerodynamic force generation, and a further investigation of this phenomenon in the flight physics and dynamics utilizing experiment is recommended.

INDEX

'*Note*: Page numbers followed by "f" indicate figures and "t" indicate tables.'

A

Absolute nodal coordinate formulation (ANCF), 73–75
Active wing morphing bioinspired test platform, 16f
Advanced Dynamic Analysis of Mechanical Systems (ADAMS), 28–31
Aerial vehicles, 1–4
Aerodyamics model, 157–158
Aerodynamic forces, 151–153, 152f
Aerodynamic loads, 76
Aerodynamic research, 14–15
Aeroelasticity, 157–158
AEROIEC. *See* Aero Load Initialized Experimental Coupled (AEROIEC)
AEROLEF. *See* Aero Load Experimental Free Flight (AEROLEF)
Aero-load blade element identification, 116–117, 118t
Aero Load Experimental Free Flight (AEROLEF), 109–112
Aero Load Initialized Experimental Coupled (AEROIEC), 109–112
Aero-model A, 112–113
Aero-model B/C, 113–117
Air Force Research Laboratory's (AFRL), 63–64
Angular velocity, 83–85, 97
ANSYS and ABAQUS, 28–31
Articulated rigid-body system, 82–85, 82f
Avian flight robotic system, 144–145, 145f
Avian flyers, 14–15
Avian scale aerodynamics, 37–44
Avian scale flapping wing flyers, 161
Avian scale ornithopter, 35–36

B

Baseline models, 14
Bat Bot, 17–18
Bioinspired flapping wing test platform
  aircraft specifications, 55, 55t
  details, 51–57
  fuselage, 53, 54f
  model verification robotic system, 57–72
    clamped test experiment–E1, 57–61, 57f–58f
    experiment and integrated forces–E1, 58–61, 59f–61f
    free-flight experiment–E2, 63–69, 64f–66f
    integrated aerodynamic force–E1-I, 63, 63f
    integrated inertial forces–E3, 71–72, 72f
    system-ID experiment–E-1, 61–63, 62f
    total forces–E2, 68–69, 70f
    vacuum camber experiment–E3, 71–72, 71f
    wing kinematics–E-1, 61, 62f
    wing kinematics–E-2, 66–69, 67f–70f
  modified ornithopter Shawn Kinkade, 51, 52f
  Morpheus Lab custombuilt test ornithopter (ML101), 51
  ornithopter structures
    material and dimensions, 53, 55t
  self-activated flap region, 53
  stiffening carbon fiber spars, 51, 52f, 53
  tail, 53, 54f
  thrust flap region, 53
  upstroke/downstroke sequence, 56, 56f
  upstroke/downstroke transition, 55–56, 56f
  wing structure bioinspired ornithopter test platform, 51, 52f
Bioinspired ornithopter test platform, 133f
Bioinspired test platform wing morphing, 16f

Blade element grid schematic aero-model A, 111f
Blade element grid schematic aero-model B, 111f
Blade element theory (BET), 38–40

C

Camber forming flexibility, 14
Camber-line/chord-line data points, 117
CAMRAD, 28–31
Cartesian velocities, 83, 85
Clamped test experiment–E1, 57–61, 57f–58f
COM-Bat, 17–18
Computational fluid dynamics (CFD), 28, 40–41, 42f
Computer-aided design (CAD), 28
COMSOL, 28–31
Constraint integrated forces, 58
Constraint model verification
  Avian flight robotic system, 144–145, 145f
  wing tip kinematics, 145–147, 146f
Constraint modes, 100
Coordinate variations, 129–132, 129f–130f
Craig–Brampton method, 74, 77–78, 99–101

D

Deformation vector, 90
Degrees of freedom (DoFs), 26–28, 58, 73
DeLaurier's model, 37–38, 113–115
Differential and algebraic equations (DAEs), 26–28
Distributed aerodynamic loads, 122–124

E

Elastic deformations, 133
Equation of motion (EOM), 26–28, 32–35, 74–75, 77, 86–91, 121
Euler angles, 88

F

Finite element method (FEM), 26–28, 91, 122
Fixed-boundary normal modes, 100
Flapping angle (FA), 55–56, 130–131
Flapping wing flight aerodynamics, 7–14, 9f
Flapping wing flight technology
  aerial vehicles, 1–4
  aerodynamic research, 14–15
  avian flyers, 14–15
  baseline models, 14
  Bat Bot, 17–18
  bioinspired test platform wing morphing, 16f
  camber forming flexibility, 14
  COM-Bat, 17–18
  first ornithopter vehicle designs, 17f
  flapping wing vehicles, 2–3
  flexible multi-body dynamics (FMBD), 18–19
  flexible multi-body structural dynamic model considering aeroelasticity (FMBSDA), 18
  flight capability, 2f
  historical considerations, 14–18, 15t, 16f–17f
  larger-scale dinosaurs, 14–15
  larger-scale natural flapping wing flyers, 14
  microaerial vehicles (MAVs), 1–2
  Microbat, 17–18
  natural flapping wing flyers, 1–4, 2f
  natural flyers, 15–16, 15t
  ornithopter performance optimization, 4–14
    desired improvements, 6–7, 6f–7f
    flapping wing flight aerodynamics, 7–14, 9f
    input variables, 5, 5f
    motion profile, 10–11, 11f
    wing flexibility, 12–13
    wing gates, 10–11, 11f
    wing geometry, 13–14
  passive wing morphing, 16f
  Phoenix, 17–18
  Reynolds number, 14–15
  rigid-multi-body-dynamics models, 18
  Robo Bats, 17–18
  smaller natural flapping wing flyers, 14
  small unmanned air vehicles (SUAVs), 1–3

Flexible multi-body dynamics (FMBD), 18–19
absolute nodal coordinate formulation (ANCF), 73–75
articulated rigid-body system, 82–85, 82f
angular velocity, 83–85
kinematic relations, 83–85
linear velocity, 85
Craig–Brampton method, 74, 77–78
definition, 73
degrees of freedom (DOFs), 73
equation of motion (EOM), 74–75, 77, 86–91
flexible body, 90, 90f
kinetic energy formulation, 86
modal superposition, 91
position vector, 89–90, 89f
potential energy formulation, 86–88
rigid-body, 89, 89f
five-body multi-body dynamics model, 79–80, 80f
angular velocity, 97
constraint modes, 100
Craig–Brampton method, 99–101
fixed-boundary normal modes, 100
inertia invariants in mass matrix, 98–99
modal synthesis method, 101–103
model generalized coordinate vector, 91–93, 92f
model orientation, 95–96
mode shape orthonormalization, 101–103
position vector, 93–95, 94f
velocity, 95
floating frame of reference (FFR), 73, 75, 78–79
Jacobian matrix, 75
linear elastic multi-body systems, 74–79
linear theory of elastodynamics (LTED), 73–75
modal neutral file (MNF), 106
relevant coordinate systems, 80, 81f
time-varying mass matrix, 78
wings structural dynamics model, 103–106, 103f, 104t, 105f–106f
Flexible multi-body dynamics model (FMBDM)
aerodynamic model, 111f, 114–116, 115f
implementation, 117–124, 120f
workflow, 122f
aerodynamic model versions formulated, 109–112, 110t, 111f
aero-load blade element identification, 116–117, 118t
Aero Load Experimental Free Flight (AEROLEF), 109–112
Aero Load Initialized Experimental Coupled (AEROIEC), 109–112
aero-model A, 112–113
aero-model B/C, 113–117
distributed aerodynamic loads, 122–124
generalized force, 124–126
generalized modal force, 126–127
generalized moments, 126
Flexible multi-body structural dynamic model considering aeroelasticity (FMBSDA), 18
Floating frame of reference (FFR), 73, 78–79
FLUENT, 28–31
Fluide structure interaction (FSI) model, 28–31, 113–114
FMBSDA. *See* Flexible multi-body structural dynamic model considering aeroelasticity (FMBSDA)
Fore-and-aft blade element approach, 116
Free-flight experiment—E2, 63–69, 64f–66f
Free-flight test analysis, 116–117
Free-flight velocity data, 117–119
Fuselage (abdominal) body dynamics, 34
Fuselage fixed reference, 148

G

Generalized force, 124–127
Generalized moment, 126
Graphical User Interface (GUI), 28–31
Grauer's model, 63

H

Harmon's blade element theory, 112
Horizontal propulsive force (HPF), 58–59

I

Individual model verification, 144
Inertia forces, 148–149
Inertia invariants, 98–99, 150–151
Instantaneous velocity, 94
Integrated aerodynamic force–E1-I, 63, 63f
Integrated forces, 147–148, 148f
Integrated inertial forces–E3, 71–72, 72f

J

Jacobian matrix, 75

K

Kinematic relations, 83–85
Kinetic energy formulation, 86

L

Large global deformations, 132–133, 133f
Larger-scale dinosaurs, 14–15
Larger-scale natural flapping wing flyers, 14
Leading edge spar flexibility, 141–143, 142f
Linear elastic multi-body systems, 24, 24f, 74–79
Linear multi-body system (LMBS) model, 26
Linear theory of elastodynamics (LTED), 73–75
Linear time invariant (LTI), 34–35
Linear velocity, 85
LMBS model. *See* Linear multi-body system (LMBS) model
Local circulatory lift force, 112–113

M

MacNeal-Schwendler Corporation (MSC), 28–31
MATLAB, 28–31
Mean absolute vertical propulsive force, 59–61
Mean deformation, 132
Mean horizontal propulsive force (MHPF), 59–61
Microaerial vehicles (MAVs), 1–2
Microbat, 17–18
Minimal average deformation, 137
Modal force, 124
Modal neutral file (MNF), 106
Modal superposition, 91
Modal synthesis method, 101–103
Modal truncation, 91
Modeled thrust flap reference motion, 140
Model generalized coordinate vector, 91–93, 92f
Modeling methodology implementation
  constraint model verification
    Avian flight robotic system, 144–145, 145f
    wing tip kinematics, 145–147, 146f
  modeling assumptions and wing flexibility
    coordinate variations, 129–132, 129f–130f
    large global deformations, 132–133, 133f
    leading edge spar flexibility, 141–143, 142f
    model input, 143, 143f
    model output, 143–144, 143f
    quantification wing flexibility, 131–132, 132f
    wing fixed reference frames, 133–140, 134f–141f
    zero deflection reference plane, 130–131, 131f
  unconstraint model verification, 147–153
    aerodynamic forces, 151–153, 152f
    inertia invariants, 150–151
    integrated forces, 147–148, 148f
    wing inertia forces, 148–150, 149f–150f
Model input, 143, 143f
Model orientation, 95–96
Model output, 143–144, 143f
Model verification robotic system, 57–72
  clamped test experiment–E1, 57–61, 57f–58f

experiment and integrated forces—E1, 58—61, 59f—61f
free-flight experiment—E2, 63—69, 64f—66f
integrated aerodynamic force—E1-I, 63, 63f
integrated inertial forces—E3, 71—72, 72f
system-ID experiment—E-1, 61—63, 62f
total forces—E2, 68—69, 70f
vacuum camber experiment—E3, 71—72, 71f
wing kinematics—E-1, 61, 62f
wing kinematics—E-2, 66—69, 67f—70f
Modern modeling methodology
aerodyamics model, 157—158
aeroelasticity, 157—158
avian scale flapping wing flyers, 161
development, 155—156
flight physics/dynamics, 158—159
implementation, 155—156
novel contributions, 158—160
ornithopter vehicle dynamics model, 156—157
scope and contributions, 156—159
Mode shape orthonormalization, 101—103
Morpheus Lab custombuilt test ornithopter (ML101), 51
MSK004 wings, 57, 57f
Multi-body dynamics (MBD)
Advanced Dynamic Analysis of Mechanical Systems (ADAMS), 28—31
angular momentum effects, 33—34
ANSYS and ABAQUS, 28—31
avian scale aerodynamics, 37—44
avian scale ornithopter, 35—36
blade element theory (BET), 38—40
CAMRAD, 28—31
code and software implementation, 28—32, 29t
computational fluid dynamics (CFD), 40—41, 42f
computer-aided design (CAD), 28
computer-aided fluid dynamics (CFD), 28
COMSOL, 28—31
degrees of freedom (DOFs), 26—28
DeLaurier's model, 37—38
differential and algebraic equations (DAEs), 26—28
dynamic frequencies, 33—34
elastic displacements, 36
equations of motion (EOMs), 26—28, 32—35
finite element method (FEM), 26—28
flapping wing vehicles aeroelastic analysis, 35—37
FLUENT, 28—31
fluidestructure interaction (FSI), 28—31
fuselage (abdominal) body dynamics, 34
Graphical User Interface (GUI), 28—31
kinematics model, 35
linear elastic multi-body models, 24, 24f
linear multi-body system (LMBS) model, 26
linear time invariant (LTI), 34—35
MacNeal-Schwendler Corporation (MSC), 28—31
MATLAB, 28—31
modeling, 26—28
methodology, 23—47, 44f—46f
NASTRAN, 28—31
nonlinear elastic multi-body models, 24—25, 25f
nonlinear strain—displacement relationships, 25—26
ordinary differential equations (ODEs), 26—28
ornithopter vehicle dynamics models, 32—35, 33f
partial differential equations (PDEs), 26—28
PATRAN, 28—31
PYTHON, 28—31
quasi-steady blade element model, 37, 38f
Reynolds Averaged Navier—Stokes (RANS), 40
rigid multi-body systems, 23—24, 24f
Ritz—Galerkin method, 26—28
Ritz method, 26—28
Schistocerca gregaria, 34—35

Multi-body dynamics (MBD) (*Continued*)
SIMPACK, 28–31
SimXpert, 28–31, 31f–32f
University of Maryland unsteady NaviereStokes solver (UMTURNS), 41–42
Vortex lattice methods (VLMs), 38–40

N
NASTRAN, 28–31
Natural flapping wing flyers, 1–4, 2f
Natural flyers, 15–16, 15t
Nodal force vector, 123
Noncirculatory force, 112–113
Nonlinear elastic multi-body models, 24–25, 25f
Nonlinear strain–displacement relationships, 25–26
North, east, down (NED) convention, 66

O
Ordinary differential equations (ODEs), 26–28
Ornithopter vehicle dynamics model, 156–157

P
Partial differential equations (PDEs), 26–28
PATRAN, 28–31
Position vector, 89–90, 89f, 93–95, 94f
Potential energy formulation, 86–88
PYTHON, 28–31

Q
Quantification wing flexibility, 131–132, 132f
Quasi-steady blade element model, 37, 38f

R
RANS. *See* Reynolds Averaged Navier–Stokes (RANS)
Reference flapping, 134
Relevant coordinate systems, 80, 81f
Reynolds Averaged Navier–Stokes (RANS), 40
Rigid multi-body systems, 23–24, 24f
Ritz–Galerkin method, 26–28
Ritz method, 26–28

S
Schistocerca gregaria, 34–35
Self-activated flap region, 53
SIMPACK, 28–31
SimXpert, 28–31, 31f–32f, 106
Small Unmanned Aerial Systems (sUAS) laboratory, 63–64
Standard finite element techniques, 76
Stiffening carbon fiber spars, 51, 52f, 53
System-ID experiment–E-1, 61–63, 62f

T
Three body-rigid body dynamics model, 61–63
Thrust flap region, 53, 111f, 112, 133
Time-variant aerodynamic forces, 116
Time-variant forcing functions, 113
Time-varying mass matrix, 78
Trailing edge flap, 116
Two-body wing simulation, 148–149

U
UMTURNS. *See* University of Maryland unsteady NaviereStokes solver (UMTURNS)
Unconstraint model verification, 147–153
aerodynamic forces, 151–153, 152f
inertia invariants, 150–151
integrated forces, 147–148, 148f
wing inertia forces, 148–150, 149f–150f
University of Maryland unsteady NaviereStokes solver (UMTURNS), 41–42
Upstroke/downstroke sequence, 56, 56f
Upstroke/downstroke transition, 55–56, 56f

V
Vacuum camber experiment–E3, 71–72, 71f
Velocity, 95
Vertical/horizontal forces, 147
Vertical propulsive force (VPF), 58–59

Vicon Vision data collection system., 57
Vicon Vision Nexus software, 65
Vortex lattice methods (VLMs), 38–40

W

Wing fixed reference frames, 133–140, 134f–141f
Wing flexibility, 12–13
Wing gates, 10–11, 11f
Wing geometry, 13–14
Wing inertia forces, 148–150, 149f–150f
Wing kinematics—E-1, 61, 62f
Wing kinematics—E-2, 66–69, 67f–70f
Wings structural dynamics model, 103–106, 103f, 104t, 105f–106f
Wing tip kinematics, 145–147, 146f
Wright-Patterson Air Force Base (WPAFB), 63–64

Z

Zero deflection reference plane, 130–131, 131f

FSC
www.fsc.org
MIX
Papier | Fördert gute Waldnutzung
FSC® C083411

Kolibri 360 GmbH
Ferdinand-Jühlke-Straße 7
99095 Erfurt, Deutschland
produktsicherheit@kolibri360.de